FLORA OF THE GUIANAS

Edited by

T0136693

S. Mota de Oliveira

Series E: Fungi and Lichens

Fascicle 3

CLADONIACEAE
(T. Ahti & H.J.M. Sipman)

2013
Royal Botanic Gardens, Kew

© Flora of the Guianas Consortium
Print edition: © The Board of Trustees of The Royal Botanic Gardens, Kew 2013

All rights reserved. No part of this publication may be reproduced, stored in a retrieval system, or transmitted, in any form, or by any means, electronic, mechanical, photocopying, recording or otherwise, without written permission of the publisher unless in accordance with the provisions of the Copyright Designs and Patents Act 1988.

Great care has been taken to maintain the accuracy of the information contained in this work. However, neither the publisher nor the editors can be held responsible for any consequences arising from use of the information contained herein.

First published in 2013 by
Royal Botanic Gardens, Kew,
Richmond, Surrey, TW9 3AB, UK
www.kew.org

Distributed on behalf of the Royal Botanic Gardens, Kew in North America by the University of Chicago Press, 1427 East 60th Street, Chicago, IL 60637, USA

ISBN 978-1-84246-479-3

British Library Cataloguing in Publication Data
A catalogue record for this book is available from the British Library

Typesetting by Margaret Newman
Publishing, Design and Photography
Royal Botanic Gardens, Kew

Printed in the UK by Marston Book Services Ltd
Printed in the USA by The University of Chicago Press

For information or to purchase all Kew titles please visit
www.kewbooks.com or email publishing@kew.org

Kew's mission is to inspire and deliver science-based plant conservation worldwide, enhancing the quality of life.

Kew receives half of its running costs from Government through the Department for Environment, Food and Rural Affairs (Defra). All other funding needed to support Kew's vital work comes from members, foundations, donors and commercial activities including book sales.

Contents

The Flora of the Guianas

is a modern, critical and illustrated Flora of Guyana, Suriname, and French Guiana designed to treat Phanerogams as well as Cryptogams of the area.

Contents: Publication takes place in fascicles, each treating a single family, or a group of related families, in the following series: A: Phanerogams; B: Ferns and Fern allies; C: Bryophytes; D: Algae; and E: Fungi and Lichens. A list of numbered families in taxonomic order has been established for the Phanerogams.
Publication of fascicles will take place when available.
In the Supplementary series other relevant information concerning the plant collections from the Guianas appears, like indexes of plant collectors.

The Flora will, in general, follow the format of other Floras such as the *Flora of Ecuador* and *Flora Neotropica*. The treatments will provide fundamental and applied information; it will cover, when possible, wood anatomy, chemical analysis, economic uses, vernacular names, and data on endangered species.

ORGANIZATION: The Flora is a co-operative project of: Botanischer Garten und Botanisches Museum Berlin-Dahlem, *Berlin*; Institut de Recherche pour le Développement, IRD, Centre de Cayenne, *Cayenne*; Department of Biology, University of Guyana, *Georgetown*; Herbarium, Royal Botanic Gardens, *Kew*; New York Botanical Garden, *New York*; Nationaal Herbarium Suriname, *Paramaribo*; Muséum National d'Histoire Naturelle, *Paris*; Nationaal Herbarium Nederland, Naturalis Biodiversity Center, *Leiden*, and Department of Botany, Smithsonian Institution, *Washington, D.C.*

The Flora is edited by the Advisory Board: Executive Editors: S. MOTA DE OLIVEIRA, M.J. JANSEN-JACOBS, Leiden. Members: NILS KÖSTER, Berlin; P.G. DELPRETE, Cayenne; P. DA SILVA, Georgetown; E. LUCAS, Kew; T.R. VAN ANDEL, Leiden; B. TORKE, New York; D. TRAAG, Paramaribo; O. PONCY, Paris, and P. ACEVEDO RODRÍGUEZ, Washington, D.C.

PUBLICATION: The *Flora of the Guianas* is a publication of The Royal Botanic Gardens, Kew. The price of the fascicles will be determined by their size. Authors are requested to turn in a hard copy of their manuscript, as well as the manuscript in an electronic version; WORD and other major systems are acceptable.

INFORMATION: http://www.nationaalherbarium.nl/FoGWebsite/index.html

Editorial office (for correspondence on contributions, etc.):
S. Mota de Oliveira
M.J. Jansen-Jacobs
Nationaal Herbarium Nederlandr
Leiden University branch
P.O. Box 9514
2300 RA Leiden
The Netherlands
Email: sylvia.motadeoliveira@ncbnaturalis.nl

CLADONIACEAE

by

Teuvo Ahti [1] & Harrie J.M. Sipman [2]

INTRODUCTION

The CLADONIACEAE are probably the most well-known family of lichenized fungi ("lichens"). They make unusual shrubby structures which form often extensive carpets several centimeters thick on poor and periodically humid soil, in particular in the boreal to subarctic zone. Here they are of economic importance as food for reindeer/caribu herds. Less well-known is that the family is also very diverse in the tropics, not only in the cool climate of the high mountains but also on the rather low, ancient sandstone massives in Southeastern Brazil and in the Guayana Highlands. After the completion of a monograph of the family for the Neotropics by the first author (Ahti 2000), a treatment for the Flora of the Guianas seemed an easy sideproduct, but in fact it turned out that the diversity of the family is still much underestimated. Among the 56 species treated here, 10 had to be described as new (Ahti & Sipman 2013).

In this treatment, species known from the Venezuelan part of the Guayana Highlands, i.e. recorded in the Venezuelan state Bolívar, have been included, because these are likely to occur also in the adjacent Guyana. Reports from Mt. Roraima that are probably from the Brazilian side of the mountain (Steyermark 1981) have also been treated here. Therefore, 9 out of the 56 species treated are not numbered because there were no records of their presence in the Guianas until the writing of this manuscript. In Venezuelan Amazonas additional species occur, but since they have their main distribution in the Andes, they were not considered part of the lichen flora of the Guayana Highlands s. str., viz. *Cladonia dactylota* Tuck., *C. pyxidata* (L.) Hoffm. and *C. squamosa* Hoffm.

CLADONIACEAE Zenker in Goebel, Pharmac. Waarenk. 1(3): 124. 1827-1829 (as 'Cladoniae').
Type: Cladonia P. Browne

Thallus dimorphic, with a squamulose, more rarely crustose, papillose, granulose or foliose primary thallus (thallus horizontalis), and fruticose, unbranched to intricately branched, generative podetia or vegetative

[1] Botanical Museum, P.O. Box 7, Fl-00014 Helsinki University, Finland.
[2] Freie Universitaet, Botanischer Garden & Botanisches Museum, Koenigin-Luise-Strasse 6-8, D-14195 Berlin, Germany.

pseudopodetia (thallus verticalis); fruticose structures may be lacking (ascomata sessile) or unknown; primary thallus may be soon evanescent, with podetia then functioning as primary thallus, e.g. by being provided with green algae; podetia often bearing podetial squamules, which are morphologically similar to the primary thallus; outer or upper surface corticate or ecorticate, often sorediate, usually with diffuse soralia. Conidiomata produced by most species, sessile to stipitate, either on primary squamules or on tips of podetia; in general dolioliform to pyriform; conidiophores unbranched or slightly branched, producing terminal, elongate, curved conidia. Ascomata stipitate to sessile, flat to globular apothecia, usually forming clusters of hymenial discs at tips of thallus verticalis, rarely on thallus horizontalis; disc blackish to dark or pale brown, ochraceous or red; hymenium containing highly amyloid gelatine; asci rather short and broad, cylindrical to clavate, distinctly shorter than the hymenium thickness; ascus tip structure is a variant of the *Lecanora* type, with massive, strongly amyloid apical dome, ascus wall non-amyloid except for the outer, amyloid layer; the very narrow, pin-like, slightly amyloid central zone of the apical dome is surrounded by a tube-like, strongly amyloid zone (see Honegger 1983: fig. 6); dehiscence taking place through a rostrate beak formed by the dome; paraphyses branching or not, tips capitate or not; spores usually 8 per ascus, hyaline, usually unicellular, oblong to fusiform, rarely 2-4-celled.

Photobiont usually *Asterochloris*, more rarely *Trebouxia*, producing colonies in the upper/outer medulla of the primary thallus, podetia and pseudopodetia.

Ecology: Preferably growing on oligotrophic, often rather acid substrate with sufficient water availability and high light availability, such as sandy soil, peat or organic debris in clearings in oligotrophic scrub in high rainfall areas. In addition, on thin soil over rock outcrops, rarely directly on bare rock surface, also epiphytic on lower trunks and bases of trees, and on rotten wood, especially in open woodland.

Distribution: A cosmopolitan family of c. 500 species in c. 10 genera, with its main Neotropical center in SE Brazil. Most species belong to the genus *Cladonia*.

Notes:
1. Strong hand lens (10x) or stereomicroscope are recommended for observation of the characters. Indicated colours are of fresh, dry material; herbarium specimens may become much browned. Measurements are on dry material; microscopical measurements on mounts in tap water. The photographs are from slightly pressed herbarium specimens. Many

morphological characters are transient and are visible only on podetia of the right stage. Therefore it is helpful to investigate several podetia and to pay attention to the presence of younger and older parts.

2. The treatment is primarily based on the Flora Neotropica Monograph of the family by Ahti (2000), from which the descriptions are largely taken and where additional information on distribution, synonymy, typification, chemistry and variability can be found. Additions and corrections include 10 species new to science.

3. The material includes some of the earliest reports of Cladoniaceae from the tropics, collected by Ernst Karl Rodschied in 1790, kept in GOET (Wagenitz 1982) and published by Meyer (1818).

4. Although the current species concept in Cladoniaceae tends to disregard taxa based solely on chemical characters, knowledge of the secundary chemistry is often very helpful for the identification. It is best assessed by TLC, but spot tests with certain reagents give usually sufficient indications. In this treatment the reactions with aqueous KOH solution (K), hypochlorite solution (C) and p-phenylenediamine (P) are indicated. For methods see Orange et al. (2001).

5. The group of species with crustose primary thallus, with densely branched podetia lacking cortex and scyphae, has been treated, for a while, as a separate genus *Cladina* Nyl. (Ahti 1984, 2000). However, macromolecular phylogenetic analyses have shown it to be a monophyletic group nesting within the genus *Cladonia* (Stenroos et al. 2002).

6. The numbers from the series "Lichenotheca Latinoamericana", cited under Selected specimens, refer to the exsiccata distributed by the Botanical Museum Berlin-Dahlem to several herbaria in Europe and the Americas. The cited exsiccata, with number and label information, were published in the volumes of the journal Willdenowia (see Sipman 1990, 1993, 1997).

LITERATURE

Abbayes, H. des, 1956. Quelques Cladonia (Lichens) des régions intertropicales, nouveaux ou peu connus, conservés dans l'herbier de Kew. Kew Bull. 11 (2): 259-266.

Abbayes, H. des, 1961. Cladonia (Lichens) nouveaux ou peu connus du Vénézuela et des frontieres brésiliennes voisines. Rev. Bryol. Lichénol. 30: 117-124.

Ahti, T. 1961. Taxonomic studies on reindeer lichens (Cladonia, subgenus Cladina). Ann. Bot. Soc. Zool. Bot. Fenn. Vanamo 32 (1): i-iv, 1-160.

Ahti, T. 1973. Taxonomic notes on some species of Cladonia, subsect. Unciales. Ann. Bot. Fenn. 10: 163-184.

4

Ahti, T. 1984. The status of Cladina as a genus segregated from Cladonia. In H. Hertel & F. Oberwinkler, Beiträge zur Lichenologie: Festschrift J. Poelt, Nova Hedwigia 79: 25-61.

Ahti, T. 1986. New species and nomenclatural combinations in the lichen genus Cladonia. Ann. Bot. Fenn. 23: 205-220.

Ahti, T. 1993. Names in current use in the Cladoniaceae (lichen-forming ascomycetes) in the ranks of genus to variety. Regnum Veg. 128: 58-106.

Ahti, T. 2000. Cladoniaceae. Flora Neotropica Monograph 78. Organization for Flora Neotropica and New York Botanical Garden, Bronx. 362 pp.

Ahti, T. & H.J.M. Sipman. 2013. Ten new species of Cladonia (Cladoniaceae, Lichenized Fungi) from the Guianas and Venezuela, South America. Phytotaxa 93: 25-39.

Ahti, T. & S. Stenroos. 1986. A revision of Cladonia sect. Cocciferae in the Venezuelan Andes. Ann. Bot. Fenn. 23: 229-238.

Evans, A.W. 1947. A study of certain North American Cladoniae. Bryologist 50: 14-51.

Evans, A.W. 1955. Notes on North American Cladoniae. Bryologist 58: 94-112.

Filson, R.B. 1981. A revision of the lichen genus Cladia Nyl. J. Hattori Bot. Lab. 49: 1-75.

Hekking, W.H.A. & H.J.M. Sipman 1988. The lichens reported from the Guianas before 1987. Willdenowia 17: 193-228.

Honegger, R. 1983. The ascus apex in lichenized fungi IV. Baeomyces and Icmadophila in comparison with Cladonia (Lecanorales) and the non-lichenized Leotia (Helotiales). Lichenologist 15: 57-71.

Hue, A.M. 1898. Lichenes extra-europaei a pluribus collectoribus ad Museum Parisiense missi I. Nouv. Arch. Mus. Hist. Nat., sér. 3, 10: 213-280.

Huovinen, K. & T. Ahti. 1986. The composition and contents of aromatic lichen substances in Cladonia section Unciales. Ann. Bot. Fenn. 23: 173-188.

Jørgensen, P.M., P.W. James & C.E. Jarvis. 1994. Linnaean lichen names and their typification. Bot. J. Linn. Soc. 115: 261-405.

Meyer, G.F.W. 1818. Primitiae Florae Essequeboensis adjectis descriptionibus centum circiter stirpium novarum; observationibus criticis (Lichens on pp. 295-298). Göttingen.

Orange, A., P.W. James & F.J. White. 2001. Microchemical methods for the identification of lichens. British Lichen Society, London.

Sandstede, H. 1932. Cladoniaceae I. In E. Hannig & H. Winkler, Die Pflanzenareale 3(6): 63-71, Karte 51-60. Jena.

Schomburgk, M.R. 1848 (1849). Versuch einer Fauna und Flora von Britisch Guiana. In R.H. Schomburgk, Reisen in Britisch Guiana in den Jahren 1840-1841, Th. 3 (Lichens: II. Region des Urwaldes: 861-867; III. Region der Sandsteinformation: 1041). Leipzig.

Sipman, H. J. M. 1990. Lichenotheca Latinoamericana a museo botanico berolinensi edita, fasciculum primum. Willdenowia 19: 543-551.

Sipman, H. J. M. 1993. Lichenotheca Latinoamericana a museo botanico berolinensi edita, fasciculum secundum. Willdenowia 23: 305-314.

Sipman, H. J. M. 1997. Lichenotheca Latinoamericana a museo botanico berolinensi edita, fasciculum tertium. Willdenowia 27: 273-280.

Sipman, H. J. M. 2007. Conservation aspects of the lichen genus Cladonia in the Guianas. Flora of the Guianas Newsletter 15: 70-72.

Stenroos, S. 1986. The family Cladoniaceae in Melanesia. 2. Cladonia section Cocciferae. Ann. Bot. Fenn. 23: 239-250.

Stenroos, S. 1989. Taxonomic revision of the Cladonia miniata group. Ann. Bot. Fenn. 26: 237-261.

Stenroos, S., J. Hyvönen, L. Myllys, A. Thell & T. Ahti. 2002. Phylogeny of the genus Cladonia s. lat. (Cladoniaceae, Ascomycetes) inferred from molecular, morphological, and chemical data. Cladistics 18: 237-278.

Steyermark, J.A. 1981. Erroneous citations of Venezuelan localities. Taxon 30: 816-817.

Vainio (Wainio), E.A. 1897. Monographia Cladoniarum Universalis I. Acta Soc. Fauna Fl. Fenn. 4: 1-509.

Vainio (Wainio), E.A. 1894. Monographia Cladoniarum Universalis II. Acta Soc. Fauna Fl. Fenn. 10: 1-499.

Vainio (Wainio), E.A. 1897. Monographia Cladoniarum Universalis III. Acta Soc. Fauna Fl. Fenn. 14 (1): 1-268.

Wagenitz, G. 1982. Index collectorum principalium Herbarii Gottingensis. Systematisch-Geobotanisches Institut der Georg-August-Universität Göttingen.

KEY TO THE GENERA

1 Fruticose parts (pseudopodetia) with thick, horny cortex and thin, arachnoid medulla, without stereome . . . . *1. Cladia*
   Fruticose parts (podetia) with thin or without cortex, usually with thick, conglutinated central cylinder (stereome), otherwise fruticose parts absent
   . . . . *2. Cladonia*

1. **CLADIA** Nyl., Bull. Soc. Linn. Normandie, sér. 2, 4: 167. 1870. Type, designated by Filson 1981: C. aggregata (Sw.) Nyl. (Lichen aggregatus Sw.)

Primary thallus consisting of small, often insignificant or evanescent basal squamules or granules from which fruticose pseudopodetia arise, more rarely the thallus mainly foliose, with very short fruticose parts. Pseudopodetia usually erect, branching, usually with numerous lateral perforations; colour ranging from black through dark brown to greenish grey; squamules absent; cortex well developed, composed of conglutinate, longitudinally arranged hyphae; medulla usually thin, arachnoid, without stereome; central canal present, but sometimes partly filled with very loose hyphal tissue (inner medulla). Conidiomata terminal on sterile pseudopodetia, containing hyaline slime; conidia colourless, elongate, curved. Ascomata shortly stalked, apothecia at tips of pseudopodetia, often aggregated, peltate, lecideine; disc brown to black; margin present only in young ascomata; spores simple, 8 per ascus, long-ellipsoid to subfusiform. Chemistry: common secondary substances include usnic acid, divaricatic acid, atranorin, fumarprotocetraric acid, protocetraric acid and ursolic acid; in *C. aggregata* several other compounds may occur, notably barbatic acid.

Distribution and ecology: *Cladia* is a genus of 12 species, most of them being confined to New Zealand, Tasmania and SE Australia; 4 occur in S America. They are most frequent in boggy, peaty localities, but also occur on soil in more or less open habitats, and occasionally on mossy trees. In the Guianas, 2 species were found, mainly in the Pakaraima Mts. of Guyana.

KEY TO THE SPECIES

1   Forming flattened mats; branching looser, often anisotomous, with distinct
    main stems . . . . *1. C. aggregata*
    Forming semiglobose heads; branching extremely dense, isotomous, without
    distinct main stems . . . . *2. C. globosa*

1. **Cladia aggregata** (Sw.) Nyl., Bull. Lich. Normandie, sér. 2, 4: 167. 1870. – *Lichen aggregatus* Sw., Prodr. 147. 1788. – *Cladonia aggregata* (Sw.) Spreng., Syst. Veget. 4(1): 270. 1827. – *Clathrina aggregata* (Sw.) Müll. Arg., Flora 66: 80. 1883. Type: Jamaica, Swartz s.n. (lectotype S, designated by Filson 1981, probable isolectotypes BM, BM-ACH 748, G, H-ACH 1583, PC-Hue, UPS-Thunberg).
                                – Fig. 1

Pseudopodetia forming dense tufts or entangling among other lichens and bryophytes, dark brown to blackish in full sunlight, pale yellowish to straw-coloured, or pale green, in shade, 0.5-10(-15) cm tall, richly branched, branching irregularly anisotomously dichotomous; thickest branches in sterile pseudopodetia 0.5-2(-8) mm diam., cylindrical to somewhat flattened or angulate, dilated at axils, horny, fragile when dry; fertile pseudopodetia much thicker (up to 2 mm diam.) and taller than sterile ones,

Fig. 1. *Cladia aggregata*; perforate pseudopodetia (Mattick 633 (B)). Bar = 2 cm.

usually more perforated and more branched towards apex. Pseudopodetial surface glossy or matt, smooth to somewhat furrowed. Pseudopodetial wall 140-180(-200) μm thick; cortex 50-90 μm, cartilaginous, perforate; perforations round to elliptic, abundant to scarce; medulla subarachnoid, 20-80(-120) μm, with 5-9 μm thick hyphae, containing scattered algal cell clusters; surface of central canal loosely arachnoid, white. Chemistry: barbatic acid, usually with traces of 4-O-demethylbarbatic acid, with or without stictic acid (TLC of 8 specimens). Colour reactions: P-, K-, KC- or P+ yellow, K+ yellow, KC-.

Distribution and ecology: An almost cosmopolitan species avoiding only the temperate to cold zones of the Northern Hemisphere and the Antarctic. Widespread in tropical America, extending from Mexico and the West Indies southwards to Chile, but absent from large areas of the tropical lowlands and from arid regions. It grows on very poor, acidic soil or mossy rocks in slightly shady places in scrub or on road banks, often on sandstone in savannas or in páramo in the mountains, in high precipitation areas, in altitude ranges from c. 300 to 4500 m alt. In the Guianas it is mainly found in the Pakaraima Mts. of Guyana, once recorded in the Bakhuis Mts. in W Suriname. Over 300 collections studied (GU: 17; SU: 1).

Selected specimens: Guyana: Pakaraima Mts., Mt. Membaru, Maas & Westra 4276 (L); Upper Mazaruni Region, Karowtipu Mt., Boom & Gopaul 7673 (NY); Potaro-Siparuni Region, Kaieteur Falls, DePriest et al. 9290 (US); Cuyuni-Mazaruni Region, N of Chimoweing Village, Pipoly 10509 (H). Suriname: Bakhuis Mts., Florschütz & Maas 2969 (L).

Note: The chemical substances found in the Guianas belong to the commonest of the species. In other areas of S America this species shows much more chemical variation: fumarprotocetraric, diffractaic, squamatic and stictic acids are among the major substances encountered (Ahti 2000).

2. **Cladia globosa** Ahti, Fl. Neotr. Monogr. 78: 39. 2000. Type: Venezuela, Bolívar, Dist. Piar, Macizo del Chimantá, NE section of Acopán-tepui, 1950 m alt., Ahti, Huber & Pipoly 45156 (holotype VEN, photo H, isotypes B, COL, DUKE, H, HOB, NY, US).

– Fig. 2

Pseudopodetia dark brown in exposed parts, green to grey in shade, up to 12 cm tall, forming continuous clumps up to 20 cm or more in diam., composed of 6-10 cm wide, semiglobose heads; branchlets cylindrical to somewhat flattened, mostly 0.5-1.5 mm thick (in fertile pseudopodetia 1-4.5 mm thick); branching type very dense isotomous dichotomy, with no main stem distinguishable, with short internodes (at the tips 3-5 internodes within 0.5 cm); axils closed. Pseudopodetial surface extremely smooth, glossy, usually without perforations on exposed parts of the pseudopodetium, with up to 10 oval holes per 1 cm unilaterally on non-exposed sides. Pseudopodetial wall horny but less fragile than in *C. aggregata*, somewhat translucent, showing

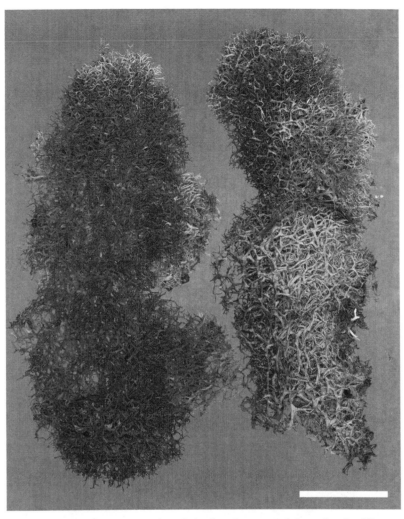

Fig. 2. *Cladia globosa*; pseudopodetia forming semiglobose heads (Sipman 39937 (B)). Bar = 2 cm.

longitudinal striations below the surface; cortex hyaline, brown at outer surface, cartilaginous, 35-55 μm thick; medulla up to 20 μm thick, white, of very loose texture, in part lacking or poorly developed, with scattered algal cell clusters below the cortex; hyphae 7-12 μm thick; central cavity wide, without inner medulla. Chemistry: barbatic acid and traces of 4-0-demethylbarbatic acid (TLC of 4 specimens). Colour reactions P-, K-, KC-.

Distribution and ecology: A Guayana Highland endemic, widespread at middle and upper elevations (700-2100 m) in SW Venezuela. It grows preferably in partial shade in scrub savanna on sandstone plateaux, often forming extensive patches mixed with tall, often semiglobose head-forming *Cladonia* species. Known in the Guianas only from a few spots in the Pakaraima Mts. of western Guyana, from c. 1000 to 1200 m alt. Over 20 collections studied (GU: 5).

Selected specimens: Guyana: Upper Mazaruni Region, Mt. Latipu, 8 km N of Kamarang, Sipman & Aptroot 19153 (B); Region 7 (Upper Mazaruni Distr.), N of Paruima Mission, Aymatoi savanna, Sipman 39937 (B, BRG).

Notes: *Cladia globosa* is characterized by its extremely densely branched cushions forming semiglobose heads. Otherwise it is practically identical with *C. aggregata*, and poorly developed plants would be indistinguishable. Well developed individuals of both species were found at short distances on the Aymatoi savanna. Here, *C. globosa* seems to prefer undisturbed places, while *C. aggregata* is the only *Cladia* present on recently burnt sites.
The phenolic chemistry of *C. globosa* is the same as in the most frequent chemotype of *C. aggregata* and is not distinctive.

2. **CLADONIA** P. Browne, Civ. Nat. Hist. Jamaica 81. 1756, *nom. cons.* Type (Jørgensen *et al.* 1994: 358): C. subulata (L.) F.H. Wigg. (*Lichen subulatus* L.)

Cladina Nyl., Not. Sällsk. Fauna Fl. Fenn. Förhandl. 8: 110. 1866 (June, preprint; '1882'). – *Cladonia* subgenus *Cladina* (Nyl.) Leight., Lich.-Fl. Gr. Britain 66. 1871. – *Cladonia* subgenus *Eucladonia* sect. *Perviae* subsect. *Cladina* (Nyl.) Mattick, Repert. Spec. Nov. Regni Veget. 49: 142. 1940 ('*Cladinae*'). Type (Ahti, 1961): *Cladina rangiferina* (L.) Nyl.

Primary thallus mostly squamulose, squamules usually elongate, in the range of 1-10 mm long and 0.5-3 mm wide, ascending, growing on one end, attached at the other (basal) end, flat, upper side greenish, lower side whitish, often resting on an extensive subterranean, mycelioid, black, brown or orange hypothallus; in many species persistent, associated with determinate growth of the podetia, otherwise evanescent, especially when podetium growth is indeterminate; in some species crustose and soon (in 1-3 years?) disappearing, consisting of grey to yellowish-grey, ecorticate areoles or verruculose granules, 0.1-0.4 mm wide; in a few species with scattered marginal, usually white, c. 1 mm long rhizomorphs. Podetia present, usually fruticose, rarely absent (hymenial discs sessile on the squamules), hollow (rarely solid), simple or branched, sometimes intricately so, usually one or a few cm tall, up to 15 cm (exceptionally 25 cm or more), often forming 10-30 cm (or more) wide colonies, sometimes forming scyphi (wine glass-like widenings, see fig. 51) or funnels (leading into the central canal, see fig. 32), or with isotomous, subisotomous or anisotomous, and dichotomous, trichotomous or tetrachotomous branching type, with or without distinct main stems; in many species of determinate growth, in others of indeterminate growth and probably persisting in living condition for decades, mainly growing at the tops, and to a lesser degree intercalarily and at lateral branch tips; basal parts gradually dying and decaying (necrotic) and discoloured, sometimes melanotic (black). Podetial surface mostly corticate, more rarely corticoid or clearly ecorticate and felty, often bearing squamules, or producing soredia (< 80 μm in diam., with felty surface) or granules (larger, with smooth surface) in more or less delimited spots, very rarely truly isidiate. Podetial wall composed of an outer cortex (sometimes absent), a looser medulla containing clusters of algal cells, a sharply delimited central cylinder, usually composed of prosoplectenchymatous, cartilaginous, rather transparent stereome (sometimes absent), surrounding a central canal; surface of the central canal usually glossy and slightly rugulose due to protruding hyphae, sometimes dull and felty or granular. Conidiomata frequent, either basal on primary thallus or terminal on tips of podetia, sometimes shortly stalked, pycnidial, containing hyaline or red slime; conidia filiform, curved, 5-14 × 0.5-1 μm. Ascomata consisting of podetia and apothecium-like hymenial discs, the latter brown, ochraceous or red, developing at tips of podetia; spores fusiform, oblong or ovoid, simple, colourless, 8 (rarely 4 or 6) per ascus, 6-24 × 2-5 μm. Chemistry: distinctive phenolic secondary compounds may be present, belonging to the depsides, depsidones, dibenzofurans and aliphatic acids. Frequent major compounds include the orcinol depside homosekikaic acid (with several related substances), the β-orcinol depsides atranorin, barbatic, squamatic and thamnolic acids, the β-orcinol depsidones fumarprotocetraric (the most frequent compound!), psoromic, stictic and norstictic acids, the dibenzofuran didymic acid and the related usnic acid, the naphthoquinone pigment rhodocladonic acid (in red hymenial discs), and the aliphatic bourgeanic, rangiformic and protolichesterinic acids.

Distribution and ecology: Worldwide, especially in humid regions, from the tropical lowlands to the maritime antarctic and middle arctic areas. The genus *Cladonia* is most abundant in humid, cool to cold climates, and absent from arid areas. Epigeic, epixylic, epidendric and epilithic, especially on acidic substrates, some species on calcareous soils. Widespread in the Guianas, with 47 species recorded, but most common in white sand or sandstone areas. Here most abundant on soil or flat rock surfaces in open vegetation. Elsewhere restricted to decaying wood and detritus.

KEY TO THE SPECIES
(incl. species from the Venezuelan Guayana Highlands)

A   Thallus squamules absent or very scarce in old thallus parts and hard to find; podetia dominant and strongly branched, never forming scyphi

1   Podetium surface without cortex, slightly felty (use strong lens!) (group *Cladina*, incl. *Cladonia signata*) 2
    Podetium surface with cortex, smooth (use strong lens!) (groups *Unciales* and *Perviae* p.p.) 11

2   Thallus colour greenish to yellowish grey; usnic acid present (KC+ yellow) 3
    Thallus colour whitish to dark grey; usnic acid absent (KC-) 4

3   Thallus forming broad, dense, semiglobose heads; trichotomous branchings often frequent; no main stems distinguishable; P- (perlatolic acid) . . . .
    . . . . *5. C. confusa*
    Thallus forming rather loose mats; trichotomous branchings always rare; main stem in part distinguishable; P+ red (fumarprotocetraric acid) . . . .
    . . . . *9. C. densissima*

4   Podetia near the tips with thick and compact outer felt layer, with plane surface without protruding algal cell clusters and without visible stereome surface; main stems always clearly distinct except near the tips; tips often deflexed 5
    Podetia near the tips with thin and arachnoid outer felt layer, with rugulose surface caused by protruding algal cell clusters and often with visible stereome surface; main stems present or absent; tips spreading to deflexed 7

5   Apical branchlets blunt with very thick felt layer; mostly over 0.5 mm wide at 1 mm below the tips; colour whitish to pale grey . . . . . . *1. C. argentea*
    Apical branchlets subulate with thinner felt layer; under 0.5 mm wide at 1 mm below the tips; colour pale grey, often with a brownish to violet tinge 6

6   Apical branchlets with mostly short (c. 0.1 mm long) brownish tips; without or with few discoloured branchlets near the tips; colour pale grey to whitish grey; common at lower to mid elevations . . . . *34. C. sprucei*
    Apical branchlets with long (c. 0.2-0.5 mm) brownish tips; often with browned branchlets near the tips; colour bluish- to violet- or brownish grey; rare high-altitude species (not yet found in the Guianas). . . . .
    . . . . *C. rangiferina* subsp. *abbayesii*

7 Stereome strongly blackening at base of podetia; top branchlets also blackening, slender, dichotomously branched, deflexed; anisotomy pronounced, with distinct main stems; heads narrow, not semiglobose (not yet found in the Guianas) . . . . *C. atrans*
   Stereome and tips not or little blackening; isotomy pronounced but main stems sometimes distinguishable in basal parts; forming broadly rounded, often semiglobose heads. 8

8 Thallus greenish or brownish grey; usually K- (atranorin absent) 9
   Thallus ashy grey; K+ yellow (atranorin present) 10

9 Thallus forming usually broad, confluent, often flattish colonies; P+ (fumarprotocetraric acid present, often with homosekikaic acid agg.); main stems absent; surface largely bare . . . . *31. C. signata*
   Thallus forming irregularly semiglobose heads; main stems distinguishable in basal parts; P- (perlatolic acid present); surface usually clearly felty; Guianas populations often forming rather separate, columnar podetia . . . .
   . . . . *5. C. confusa*

10 Thallus forming very dense, regularly semiglobose heads; main stems absent; stereome never blackening. . . . . *27. C. rotundata*
   Thallus forming less dense, irregularly semiglobose heads; main stems distinguishable in basal parts; stereome often blackening at base . . . .
   . . . . *8. C. dendroides*

11 Thallus colour greenish to yellowish grey; usnic acid present. 12
   Thallus colour whitish-grey to brown; usnic acid absent. 21

12 Main stems for the most part over (2-)5 mm thick, very strongly and irregularly branched; wall densely split and perforated except when young; P+ red, P+ pale yellow or P-, K-, K+ yellow turning orange or K+ brownish (usually containing fumarprotocetraric acid or stictic acid). . . . .
   . . . .*38. C. subreticulata*
   Main stems under 2 mm thick, less strongly and more regularly, usually dichotomously branched; wall not perforated or at axils only; wall of central canal not reticulate; usually P+ yellow or P-, K+ yellow or K- (containing thamnolic, barbatic or squamatic acids), rarely P+ red, K+ brownish . . .13

13 Central canal of podetium with glossy, smooth wall 14
   Central canal of podetium with matt, puberulent or felty wall. 18

14 Podetia mostly under 0.4 mm wide, without distinct main stems . . . .
   . . . . *19. C. peltastica*
   Podetia 0.5-1 mm wide, often with main stems 15

14

15   Moderately branched, forming loose tufts or mats with many thick main
     stems; mostly P+ yellow, K+ yellow (thamnolic acid) 16
     Densely branched, forming "spiny" heads with thin, indistinct main stems;
     mostly P-, K- (barbatic and/or squamatic acid), rarely P+ yellow or red, K+
     yellow or K- 17

16   Branchlets at the podetium tips at an obtuse angle; squamules always absent;
     widespread in sandstone tableland ....*44. C. vareschii*
     Branchlets at the podetium tips at a sharp angle; squamules sometimes
     present, scarce; mainly in *Sphagnum* bog at higher elevation (not yet found
     in the Guianas) ....*C. flavocrispata*

17   Thallus P-, K- (barbatic and/or squamatic acid) or P+ yellow, K+ yellow
     (thamnolic acid) ....*33. C. spinea*
     Thallus P+ red, K- or K+ brownish (fumarprotocetraric acid) (not yet found
     in the Guianas) ....*C. chimantae*

18   Stereome absent, replaced by a compacted but not cartilagineous medullary
     layer; P+ yellow (thamnolic acid) (not yet found in the Guianas) ....
     ....*C. crassiuscula*
     Stereome present, cartilagineous; P+ yellow, K+ yellow or P-, K- 19

19   Thin, richly branched, creeping to erect, usually without main stems; surface
     of central canal somewhat fibrose; fine needle crystals developing on
     podetial tips in herbarium material; P- or +orange, K- (fumarprotocetraric
     acid sometimes present) .... *41. C. substellata*
     Stoutish, erect, often with dominant main stems; surface of central canal
     smooth; not developing fine needle crystals on podetial tips; P+ yellow, K+
     yellow (thamnolic acid), rarely P-, K- (squamatic acid) 20

20   Podetia thick, scarcely and strongly anisotomously branched, pale greyish
     yellow with brown-variegated parts towards the base; cortex thin; stereome
     thin and soft; always P+ yellow, K+ yellow............. *42. C. sufflata*
     Podetia slender, moderately, more or less anisotomously branched, clearly
     yellow, uniformly coloured throughout; cortex thick; stereome strong; P+
     yellow, K+ yellow or P-, K- .... *35. C. steyermarkii*

21   Branching usually clearly anisotomous, distinct main stems present; not
     forming very dense, rounded heads 22
     Branching mostly isotomous, distinct main stems absent; usually forming
     very dense, rounded to elongate heads........................25

22   Podetia little branched, thick (up to 3 mm); stereome soft and white; wall of
     central canal pruinose; among mosses on peat.......... *42. C. sufflata*
     Podetia much branched, thin (up to 1 mm); stereome hard and hyaline; wall of
     central canal shiny; usually free-growing on sand or sandstone flats 23

23 Branchlets ending in fine, blackish tips (0.5-)1-2 mm long and 0.1-0.2 mm wide (not yet found in the Guianas)...................... *C. huberi*
Branchlets tips pale or shorter and wider 24

24 Podetia of uniform, grey colour; axils often closed, not much dilated; common in white-sand savannas near the coast......... *32. C. sipmanii*
Podetia variegated with whitish and brown patches, particularly on older parts; axils mostly perforated and often widely dilated and funnel-shaped; only in sandstone tablelands of the interior (not yet found in the Guianas)
.... *C. hians*

25 Stereome replaced by a layer of incompletely conglutinated hyphae, somewhat fibrous when broken; forming semiglobose heads; apical branchlets incurved.......................... *25. C. pulviniformis*
Stereome completely conglutinated, glassy 26

26 Apical branchlets mostly pointing upwards, not variegated; internodes usually over 2 mm long; hymenial disc-bearing podetia conspicuously swollen 27
Apical branchlets pointing in all directions, sometimes variegated; internodes hardly over 2 mm long; hymenial disc-bearing podetia unchanged 28

27 Podetia very thin and rugulose, usually less than 0.3 mm, whitish; only squamatic acid present .... *26. C. recta*
Podetia thicker and usually smooth, c. 0.3-0.4 mm, whitish or yellowish; chemistry various .... *19. C. peltastic*

28 Podetia 0.4-0.8 mm thick, usually variegated; K+ yellow (thamnolic acid)
.... *45. C. variegata*
Podetia 0.2-0.4(-0.8) mm thick, sometimes variegated; K- or brownish (fumarprotocetraric acid) 29

29 Apical branchlets c. 0.1-0.2 mm wide, rough by protruding clusters of algal cells... *31. C. signata*
Apical branchlets c. 0.4 mm wide, smooth................ *15. C. maasii*

B   Thallus squamules present, often abundant and dominating; podetia usually scarcely branched, sometimes forming scyphi, in some species richly branched or absent

1 Mature thallus dominated by primary thallus squamules; podetia also in adult stage (with hymenial discs) scarcely longer than the thallus squamules 2
Mature thallus dominated by podetia, which exceed the squamules in length many times 11

2   Podetia present, beset with dehiscent, recurved squamules easily falling off
    and scarce in herbarium specimens; rarely with (brown) hymenial discs; P+
    red (fumarprotocetraric acid) 3
    Podetia absent or present, without or with non-dehiscent squamules, often
    with red hymenial discs; P- or P+ yellow (when P+ red, go to 3) 4

3   Dehiscent podetial squamules rounded; sterile podetia obtuse, occasionally
    with very narrow scyphi; basal squamules without marginal fibrils . . . . . .
    . . . . *21. C. pityrophylla*
    Dehiscent podetial squamules narrowly elongate, often almost isidia-like;
    sterile podetia subulate, never with scyphi; basal squamules usually with
    scattered, white marginal fibrils . . . . *4. C. ceratophylla*

4   Medulla and lower side of squamules red . . . . *17. C. miniata*
    Medulla and lower side of squamules white or pale brownish, or with vein-
    like ochraceous stripe 5

5   Basal squamules without soredia or isidia, well developed and deeply lobed,
    up to over 5 mm long; podetia usually over 5 mm tall. . . *30. C. secundana*
    Basal squamules with soredia or isidia 6

6   Basal squamules with short cylindrical isidia on the margins, without
    soredia . . . . *14. C. isidiifera*
    Basal squamules without isidia, with sorediate margins 7

7   Basal and podetial squamules large, 10-20 mm long; margins farinosely
    sorediate; medulla white . . . . *16. C. meridionalis*
    Basal and podetial squamules small, up to 1-5 mm long; margins or lower
    side sorediate; medulla sometimes ochraceous below 8

8   Basal squamules rounded to elongate, thick (c. 0.3 mm), ascending; thallus
    appearing squamulose 9
    Basal squamules short, deeply lobed, thin (c. 0.15 mm) or thicker, much
    appressed to its substrate centrally; thallus appearing crustose 10

9   Basal squamules terminaly rounded, apically sorediate on lower side, with
    concave upper side and brownish lower side; K- . . . . . . *3. C. cayennensis*
    Basal squamules terminally deeply dissected and granular, with flat to
    convex upper side and often with an ochraceous vein-like stripe below; K+
    yellow . . . . *13. C. hypoxantha*

10 On termite mounds; K+ brownish, P+ red (fumarprotocetraric acid) . . . . . .
    . . . . *43. C. termitarum*
    On tree bark; K+ yellow, P+ yellow (thamnolic acid) . . . . . *7. C. crustacea*

11 Podetium surface largely felty and without cortex, or with smooth, denudated stereome, more or less covered with deciduous soredia, granules or squamules, or strongly verrucose............................12
    Podetium surface smooth and continuously or largely corticate, without soredia or granules and not strongly verrucose, sometimes with squamules 26

12 Podetia wineglass-shaped, scyphi at least three times wider than stalk (in well-developed podetia)....................................13
    Podetia not wineglass-shaped, scyphi absent or only slightly wider than the rest of the podetium 15

13 Podetia with yellowish/bright greenish tinge (usnic acid present, KC+ yellow, P+ yellow, K+ yellow); hymenial discs red; scyphi sorediate and/or granular, stalk corticate..14
    Podetia not yellowish, pale grey (fumarprotocetraric acid present, KC-, P+ red, K-/brownish); hymenial discs brown; scyphi and stalk sorediate.....
    ....*40. C. subsquamosa*

14 Podetia farinosely sorediate .... *18. C. mollis*
    Podetia smooth to coarsely granulose, occasionally with some soredioid granules ....*6. C. corallifera*

15 Podetia with more or less deciduous small squamules or corticate granules, not truly sorediate (surface not granular or mealy), simple or little branched; tips usually persistently subulate, sometimes with narrow scyphi 16
    Podetia sorediate to granular-sorediate, often very densely so (with granular or mealy surface), branchy or not; podetia often finally with narrow scyphi 22

16 Podetia completely denudated, with horny, brownish, smooth stereome ...17
    Podetia more or less denudated, with whitish, felty surface............18

17 Podetia usually simple, usually 1-2 cm tall, soon denudated and without squamules; hymenial discs common, red.............. *10. C. didyma*
    Podetia branchy, often 3-5 cm tall, persistently squamulose; hymenial discs rare, brown .... *20. C. persphacelata*

18 P+ yellow, K+ yellow (thamnolic acid); podetia slender, with numerous squamules; without scyphi, but sometimes with funnel-shaped, wide open axils 19
    P+ red, K-/brownish (fumarprotocetraric acid); podetia slender to stout; squamulose or not; often with scyphi..........................20

19 Podetia slender, often with soredioid granules; with closed axils; on bark of living trees .... *36. C. subdelicatula*
Podetia more robust, without soredioid granules; apically with widened, funnel-shaped, open axils; on sandy soil or litter..... *23. C. polystomata*

20 Podetial squamules rounded, horizontal, often strongly concave or convex; podetia short, up to 1 cm tall, not ending in scyphi ....*21. C. pityrophylla*
Podetial squamules elongate, down-turning, flat to slightly convex; podetia about 0.8-3 cm tall, ending in scyphi or not......................21

21 Podetia usually ending in small scyphi; surface with squamules..........
....*11. C. furfuraceoides*
Podetia without scyphi, always with subulate tips; surface with granules...
.... *2. C. cartilaginea*

22 Podetia thick and short, pale whitish yellow to whitish (usnic acid often present in low amounts), densely sorediate; corticate near base; apothecia often present, red; on wood; P+ yellow, K+ yellow (thamnolic acid) .....
....*24. C. prancei*
Podetia tall subulate or branchy, grey to brownish (usnic acid never present), thinly sorediate or with granules, which may be attached to each other and form microsquamules; hymenial discs uncommon, usually brown; on wood or sand; P+ red, K- (fumarprotocetraric acid) (when P+ yellow, K+ yellow, see also *C. subdelicatula*).................................. 23

23 Podetia with open axils; P+ yellow, K+ yellow (not yet found in the Guianas)....................................... *C. granulosa*
Podetia with closed axils; P+ red or P-, K-....................... 24

24 Podetia completely ecorticate, pale greenish throughout; mainly on steep faces of wood 25
Podetia corticate near base and below scyphi, pale grey and easily browning, sometimes melanotic below; mainly on mineral soil... *22. C. polyscypha*

25 Podetia mostly unbranched; P+ red; hymenial discs brown; widespread....
.... *37. C. subradiata*
Podetia apically branched; P-; hymenial discs red; known only from Rupununi savanna .... *29. C. rupununii*

26 Podetia regularly forming scyphi; scyphi proliferating from the center (not yet found in the Guianas) .... *C. rappii*
Podetia subulate or bluntish, often branchy, sometimes with enlarged open axils (funnels), or with small scyphi proliferating from the margin.... 27

27 Thallus with yellow tinge (usnic acid present)..................... 28
Thallus without yellow tinge (usnic acid absent).................. 29

28 Podetia usually less than 0.5 mm thick, moderately to densely branching, with narrowly perforated or closed axils; often abundant at lower elevations; P+ yellow or P- .... *19. C. peltastica*
Podetia usually over 1 mm wide, with widely opened axils; rare, at high elevations (not yet found in the Guianas).............. *C. flavocrispata*

29 Podetia almost unbranched, stoutish, with strongly areolate cortex and somewhat squamulose; hymenial discs red, usually present ....
....*12. C. guianensis*
Podetia branched, more or less squamulose; hymenial discs brown, rarely present 30

30 Podetia squamulose over their full length, often partly ecorticated 31
Podetia sparingly squamulose, mainly near the base, corticated throughout; P+ yellow or red, K+ yellow or K- (thamnolic or fumarprotocetraric acid)
.... 32

31 Podetia apically with widely open, funnel-shaped axils; P+ yellow, K+ yellow (thamnolic acid) .... *23. C. polystomata*
Podetia with closed, not funnel-shaped axils; P-, K- (squamatic acid) .....
....*39. C. subsphacelata*

32 Thallus whitish-grey; podetia thin, 0.4-0.6 mm, erect, fasciculate, very fragile, scarcely branched and with closed axils; cortex often rugulose ...
....*28. C. rugulosa*
Thallus ashy grey; podetia thin or stoutish, entangled and richly branched, not very fragile; axils more or less perforated 33

33 Podetia thin, without main stems, rarely over 0.4 mm wide; axils closed ...
....*19. C. peltastica*
Podetia thick, with clear main stems c. 1 mm wide; axils perforated, often with wide opening 34

34 Not all axils perforated, especially not near tips, and the perforations small; surface almost continuously corticate and of uniform, greyish colour; common in coastal white sand savanna ....*32. C. sipmanii*
All axils perforated, gaping, often funnel-shaped; surface variegated with pale and dark patches, at least in older parts of the podetia; collected only in the sandstone tablelands (not yet found in the Guianas) ......*C. hians*

1. **Cladonia argentea** (Ahti) Ahti & DePriest, Mycotaxon 78: 501. 2001. – *Cladina argentea* Ahti, Ann. Bot. Fenn. 23: 221. 1986. Type: Venezuela, Bolívar, Dist. Piar, Macizo del Chimantá, Churí-tepui, 2250 m alt., Ahti *et al.* 44914 (holotype VEN, isotypes B, COL, H, NY, MERF, US). – Fig. 3

Fig. 3. *Cladonia argentea* (Sipman 4036 (B)). Bar = 2 cm.

Primary thallus absent. Podetia 4-12 cm tall, of indeterminate growth, greyish white, with apical branchlets concolorous until the extremities (0.1 mm), forming flat, loose colonies with rather narrow (up to 1.5 cm wide) podetial heads; branching type anisotomous dichotomy (often 100% dichotomous), rarely with some trichotomies; axils never perforated; apical branchlets thick, subobtuse or tapering, divergent, rarely unilaterally deflexed; main stems distinct below the somewhat isotomous apex, 1.2-2.5 mm thick; necrotic basal parts pale grey, stereome not melanotic. Podetial surface dense-felty, smooth. Podetial wall in main stems 360-520 μm thick; felty ectal layer very dense and thick, occupying almost half of the wall width, 140-200 μm; medulla (incl. the algal glomerules) 100-140 μm; stereome 150-200 μm ($^1/_3$ of the wall in dry state). Conidiomata containing purple slime. Hymenial discs not observed. Chemistry: atranorin and fumarprotocetraric acid, with traces of protocetraric and confumarprotocetraric acids (TLC of 4 specimens). Colour reactions: P+ orange-red, K+ yellow, KC-.

Distribution and ecology: An endemic known only from upper elevations of the Guayana Highland, where it is not uncommon in the Venezuelan part. It seems to prefer hummocks of open, acidic, ombrotrophic bogs, but it may also grow on periodically dry soil and flat rock surfaces, at 1200-2600 m alt. Found in a few spots in the Pakaraima Mts. in Guyana, in open places in scrub savanna on sandstone plateaux, growing on thin soil covers of rock surface, from 400 to 1200 m alt. Over 15 collections studied (GU: 9).

Selected specimens: Guyana: Potaro-Siparuni Region, Kaieteur Falls National Park, DePriest 9261 (H, US); Region 7 (Upper Mazaruni Distr.), N of Paruima Mission, Aymatoi savanna, Sipman 39874 (B, BRG).

Notes: *Cladonia argentea* resembles most closely *C. rangiferina*, in particular robust morphotypes, but it is readily distinguished by the absence or very sparse development of brown pigment on the podetia, even at the extreme tips. *C. argentea* remains conspicuously pale, often whitish grey, even in fully open, strongly illuminated habitats, where *C. rangiferina* gets more or less brown-pigmented. The felt layer on the podetia is thicker in *C. argentea*, causing the tips to be usually blunt. At 1 mm below the tip, the podetia of *C. argentea* measure usually over 0.5 mm in width, while in *C. rangiferina* they remain usually below 0.5 mm. Moreover the slime in the conidiomata is purple in *C. argentea*, while it is hyaline in *C. rangiferina*. Finally, the branching is more strongly dichotomous in *C. argentea*.

Cladonia atrans (Ahti) Ahti & DePriest, Mycotaxon 78: 501. 2001. – *Cladina atrans* Ahti, Fl. Neotrop. Monogr. 78: 69. 2000. Type: Venezuela, Bolívar, Macizo del Chimantá, valley between Torono-tepui and Chimantá-tepui, 2100 m alt., Ahti, Huber & Pipoly 45185 (holotype VEN, isotypes B, DUKE, H, NY, US). – Fig. 4

Primary thallus absent. Podetia 5-18 cm tall, of indeterminate growth, ashy to pale grey, the branchlets with conspicuous, 0.5-1.5 mm tall, brown to black tips, forming flat, loose colonies with rather narrow (up to 1.5 cm wide) podetial heads; branching type anisotomous dichotomy, very rarely trichotomy; axils closed or often perforated; apical tips erect or divaricate (branching angle 45-90 degrees), slightly curved; main stems slender, occasionally robust, 0.5-0.9(-2.5) mm thick, fairly distinct but with many long side branches; tips melanotic, giving the colonies a variegated colour; also basal, necrotic parts melanotic. Podetial surface with loose arachnoid cover, verruculose due to protruding algal cell clusters. Podetial wall in main stems 190-275 μm thick; ectal layer loosely arachnoid, very thin and

Fig. 4. *Cladonia atrans* (Sipman 27139 (B)). Bar = 2 cm.

in part seemingly absent, 30-50 μm; medulla 25-50 μm; stereome 170-250 μm. Conidiomata containing red slime. Hymenial discs not observed, probably brown. Chemistry: atranorin and fumarprotocetraric acid, with traces of protocetraric and confumarprotocetraric acids. Colour reactions: P+ fast orange red, K+ yellow, KC-.

Distribution and ecology: Another endemic of the Guiana Shield, not yet known from the Guianas, but locally abundant in the Venezuelan sector of the Guayana Highlands. Records are available from Chimanta, Auyantepui, Guaiquinima and Jaua in Bolivar and from Huachamacari and Marahuaca in Amazonas. The species grows usually in summit bogs of sandstone table mountains, from c. 1500 to 2500 m alt. Over 14 collections studied.

Note: *Cladina atrans* resembles most closely *C. sprucei*. It can be recognized by the melanotic tips of the branchlets and basal parts of the podetia.

2. **Cladonia cartilaginea** Müll. Arg., Flora 63: 260 (= Lichenol. Beitr. 169). 1880. Type: Venezuela, near Caracas, Ernst 3 (holotype G, photo H, isotype G). – Fig. 5

Primary thallus persistent to evanescent, consisting of 5-11 mm long, 0.5-1 mm wide, somewhat laciniate, esorediate squamules. Podetia 0.6-2(-3) cm long, 0.6-1 mm wide, usually of determinate growth, whitish-

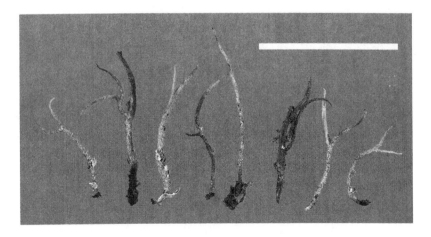

Fig. 5. *Cladonia cartilaginea* (Florschütz 176 (B)). Bar = 2 cm.

grey, occasionally blackening at the base, often markedly subulate and gradually attenuated from near the base, phyllopodial, unbranched or sparsely divaricately branched, with closed axils, not forming scyphi; sterile tips shortly subulate to blunt. Podetial surface ecorticate or at the base thinly corticate with cracks, with granules sometimes transgrading into soredia, near the tips easily becoming denudated and showing the white to pale brownish, often slightly translucid stereome. Podetial wall 130-250 μm thick; cortex absent; medulla discontinuous, 0-25 μm where present; stereome 100-250 μm, hard; surface of central canal striate. Conidiomata laminal on primary squamules or apical at podetium tips, sessile or stipitate, 180-360 μm thick, constricted at base, containing hyaline slime. Hymenial discs frequent but not constantly present; 0.3-1.5 mm wide, brown. Chemistry: fumarprotocetraric acid, usually with traces of protocetraric acid and confumarprotocetraric acid (TLC of 1 specimen). Colour reactions: P+ fast red, K-, KC-.

Distribution and ecology: The species seems widespread in the Neotropics, from the West Indies to Uruguay, but its distribution is poorly known since it is apparently undercollected. It may have a weedy character. Known so far from two samples in the Guianas, from Guyana and Suriname, from 100 to 1000 m alt., but probably overlooked. Over 50 collections studied (GU: 1; SU: 1).

Selected specimens: Guyana: Karowtipu Mt., Boom & Gopaul 7598 (NY). Suriname: Dam, Florschütz 176 (H, L).

Notes: *C. cartilaginea* is an inconspicuous and probably short-living species, perhaps most easily confused in the Guianas with *C. subdelicatula*, which differs most clearly from *C. cartilaginea* by the presence of thamnolic acid instead of fumarprotocetraric acid. *C. cartilaginea* may also resemble stages of other largely subulate *Cladoniae* from decaying organic substrate, such as *C. subradiata*, *C. polyscypha*, *C. prancei*, but these can be easily separated because they are clearly sorediate. In other cases, the chemistry will be discriminative. The species is also very similar to *C. corymbites* Nyl., erroneously reported for the Guianas (Sipman 2007), but in that species the podetia bear squamules rather than granules. Outside the Guianas, races with psoromic acid or with additionally substances of the stictic or sekikaic acid complex are known. The Suriname specimen looks slightly decayed and might be a degenerated morph of *C. polyscypha*.

Fig. 6. *Cladonia cayennensis;* type (Aptroot 15098 (L)). Bar = 1 cm.

3. **Cladonia cayennensis** Ahti & Sipman, Phytotaxa 93: 1. 2013.
Type: French Guiana, Cayenne, Botanical Garden, on palm, sea
level, 04° 56'N, 52° 20'W, March 1985, A. Aptroot 15098 (holotype
L, isotypes B, H).   – Fig. 6

Primary thallus persistent, consisting of flat to concave, even involute
squamules, 1-2 mm wide, soft and fragile, with rounded lobes, entire
or little divided; upper side greenish-brown, lower side white, floccose
and loosely sorediate, specially along margins; occasionally squamules
provided with short basal, brownish-veined stalk. Podetia, conidiomata
and hymenia discs unknown. Chemistry: usnic acid (in low concentration)
and zeorin (needle-like crystals abundant on old herbarium specimens!).
Colour reactions: P-, K-, KC-.

Distribution and ecology: Known so far only from the Guianas,
but in view of its occurrence in secondary habitats, the species is probably
more widespread. Perhaps it was once reported as *C. ahtii* from adjacent
Venezuela, in Bolívar near Canaima, epiphytic at 550 m alt.; this record
was not accepted as *C. ahtii* by Ahti (2000). The species was found
exclusively epiphytic on hepatic film on smooth, living palm stems in
plantations along the coast. Only 2 collections studied (SU: 1; FG: 1).

Additional specimen (paratype): Suriname: Paramaribo, Palmgarden, sea level, Aptroot 14829 (L).

Notes: *Cladonia cayennensis* is a sorediate representative of the *Cladonia miniata* group, similar to *C. ahtii* S. Stenroos, from which it differs by the thinner thallus lobes and different chemistry (Ahti 2000). It is also close to *C. meridionalis*, but that species has much larger thallus lobes, which are sorediate over much of their lower side. *Cladonia termitorum* is also rather similar, due to its slightly sorediate thallus lobes. It differs by its thinner and more finely divided squamules, its chemistry and the regular presence of small podetia and red hymenial discs.

In the absence of apothecia and podetia its systematic position is tentative. It is based on the similarity of its thallus squamules with the *miniata* group of the genus *Cladonia*, and its chemistry which occurs in *Cladonia* rather than in other squamulose lichen genera.

Cladonia cayennensis is easily overlooked because it is known only in squamulose, sterile state. However, having a distinct morphology and chemistry, it cannot be any other species of *Cladonia* recognized by Ahti (2000) in tropical America. Herbarium specimens are easily identified by the soredia and abundant zeorin crystals on the lower surface of the squamules (crystals expected to be absent from fresh material). The presence of a low amount of usnic acid causes hardly any yellowish colour.

4. **Cladonia ceratophylla** (Sw.) Spreng., Syst. Veg. 4(1): 271. 1827.
   – *Lichen ceratophyllus* Sw., Prodr. 147. 1788 ('*cerotophyllus*').
   Type: Jamaica, Swartz s.n. (lectotype S, designated by Ahti 1993, isolectotypes BM, G, H-ACH 1576, S, UPS-ACH).   – Fig. 7

Primary thallus persistent, forming up to 4 cm thick cushions, consisting of large, broadly laciniate squamules up to 1.5(-2.5) cm long and 2-7 mm wide, with ascending, laterally recurved lobes; branching rhizomorphs present, scattered along the margins, white to black, up to 1.3 mm long. Podetia 1-3 cm tall, 0.5-1 mm thick, of determinate growth, greenish-grey, not melanotic, simple or somewhat branched, with subulate tips. Podetial surface usually with a smoothly corticate sheath at base, ecorticate upwards; bearing dense to scattered, isidioid, recurved microsquamules or sometimes granules or soredia. Podetial wall 200-320(-440) μm thick; cortex 20-25 μm; medulla 60-100 μm; stereome 120-200(-320) μm; central canal minutely papillate. Conidiomata usually at tips of young podetia, sometimes on primary thallus, 160-260 μm thick, ampullaceous to ovoid, constricted at base; containing hyaline slime; conidia 5-7 × 1 μm slightly curved. Hymenial discs rare, at tips of usually branched, much differentiated fertile podetia, 0.5-1.5 mm wide; spores oblong, 8-13 × 2.5-

Fig. 7. *Cladonia ceratophylla* (Sipman 39947 (B)). Bar = 2 cm.

3.5 μm. Chemistry: fumarprotocetraric acid and traces of protocetraric and confumarprotocetraric acids (TLC of 2 specimens). Colour reactions: P+ orange red, K+ yellow or K-, KC-.

Distribution and ecology: Widespread in Middle and S America, with disjunct reports from La Réunion and India, often in lower mountains below 3000 m alt. Widespread in the mountainous parts of all three Guianas, from 400 to 1200 m alt., generally on soil in somewhat shady habitats in open forest. Apparently avoiding the white sand savannas of the lowlands. Over 200 collections studied (GU: 8; SU: 3; FG: 3).

Selected specimens: Exsicc.: Lichenotheca Latinoamericana 7. Guyana: W end of Kanuku Mts., Smith 3200 (G, NY, U, US); N of Kamarang, Mt. Latipu, Maas & Westra 4210 (H, L); Karowtipu Mt., Boom & Gopaul 7640 (BRG, NY). Suriname: Brownsberg, Florschütz s.n. (L); Bakhuis Mts., Florschütz & Maas 2963 (L). French Guiana: Région de l'Inini, Mont Atachi Bacca, Cremers *et al.* 10412 (B); Montagne de l'Inini, zone est, pente ouest sous le vent, Cremers *et al.* 9160 (L).

Notes: This widely distributed, common species is easily identified by its large squamules with scattered white rhizomorphs along the margins and subulate podetia with recurved, easily shed squamules. Atranorin is present as additional substance in about 50% of the specimens from outside the Guianas.

28

Cladonia chimantae Ahti, Fl. Neotrop. Monogr. 78: 321. 2000. Type:
Venezuela, Bolívar, Macizo del Chimantá, Acopán-tepui, 1950 m alt.,
Ahti, Huber & Pipoly 45161a (holotype VEN, isotypes H, NY).
– Fig. 8

Primary thallus not persistent, unknown. Podetia 4-5 cm tall, 0.5-
1 mm thick, of indeterminate growth, pale yellowish-green, necrotic
parts somewhat melanotic; branching type fairly dense subisotomous
dichotomy and trichotomy, main stems distinguishable in part, with
lowest internodes only 1-2 mm long; axils closed, rarely perforated;
branchlets divaricate, acute in main stems, attenuate and curly at tips
in lateral branches. Podetial surface verruculose, slightly arachnoid to
smooth. Podetial wall 150-175 μm thick; cortex 25 μm, rudimentary,
discontinuous; medulla 25 μm; stereome 100-130 μm, hard; central canal
finely striate. Conidiomata cylindrical, 150 × 50 μm; slime and conidia not
seen. Hymenial discs not seen. Chemistry: usnic and fumarprotocetraric
acids, with traces of protocetraric, confumarprotocetraric and (probably)
hypoprotocetraric acids and an unknown terpene. Colour reactions: P+
orange red, K-, KC+ yellow.

D i s t r i b u t i o n   a n d   e c o l o g y : A rare species from the Venezuelan part
of the Guayana Highland, known only from the type locality. Here it was
growing at 1950 m alt. over a periodically wet rock outcrop surrounded
by bogs in paramoid vegetation. Not yet known from the Guianas. Few
collections studied.

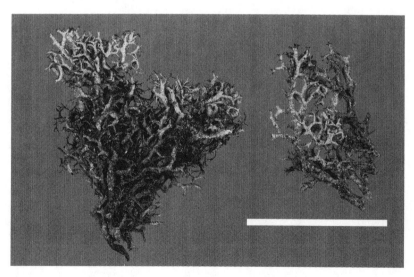

Fig. 8. *Cladonia chimantae* (Ahti et al. 45161a (VEN)). Bar = 2 cm.

Note: *Cladonia chimantae* resembles most closely *Cladonia spinea* and *Cladonia densissima*. From the former it differs most clearly by its chemistry: presence of fumarprotocetraric acid instead of barbatic or thamnolic acid. From the second it differs by the presence of some cortex and by its thicker branchlets (Ahti 2000).

5. **Cladonia confusa** R. Sant., Ark. Bot. 30A(10): 13. 1942. – *Cladina confusa* (R. Sant.) Follmann & Ahti in Follmann, Philippia 4: 321. 1981. Type: Ecuador, Imbabura, Lake Cuicocha, Islote Chica, 3150 m alt., Asplund L107, Santesson, Lich. Austroamer. 351 (holotype S, isotypes BM, G, H, S, UPS, US, W). – Fig. 9

Primary thallus absent. Podetia 6-12 cm tall, of indeterminate growth, yellowish-grey, densely branched with mature internodes 2-3 mm long, forming 2-5 cm wide, usually scattered, semiglobose heads; predominant branching type isotomous dichotomy or trichotomy; axils normally perforated; usually no main stems distinguishable; ultimate apical branchlets slender, attenuate, diverging in a wide to rather narrow angle; basal parts darkened, not necrotic. Podetial surface layer lax, thin, easily disintegrating, causing a verrucose surface in which the algal cell clusters are often visible. Podetial wall 140-200 μm thick; ectal layer 60-140 μm,

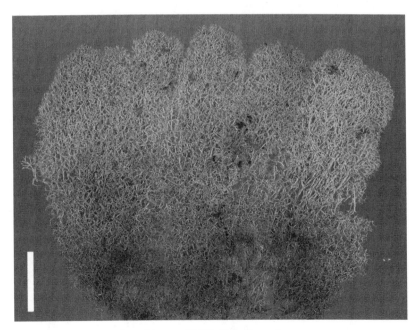

Fig. 9. *Cladonia confusa* (Sipman 39890 (B)). Bar = 2 cm.

arachnoid, not clearly differentiated in ectal layer and medulla; stereome 30-80 μm; surface of central canal minutely rugulose. Conidiomata containing hyaline slime. Hymenial discs rare, brown. Chemistry: (-)-usnic acid and perlatolic acid, with traces of anziaic acid (perhaps usually present) and several occasional, unknown substances (TLC of 8 specimens). Colour reactions: P-, K-, KC+ yellow.

Distribution and ecology: A widespread and common species in the mountains of C. and S America, where it is the most frequent species of subg. *Cladina*. Southwards, it reaches central Chile. The range of the species includes southern and East Africa, Australasia and western Melanesia. It grows on acid, poor soil and litter, in open places. In the Guianas it is a regular, but not very abundant species from the sandstone tablelands of the Pakaraima Mts., from 300 to 1000 m alt. It grows on open places in rocky shrub savannas, on thin soil on sandstone plateaux. Over 500 collections studied (GU: 15).

Selected specimens: Guyana: Pakaraima Mts., Mt. Membaru, Maas & Westra 4286a (L); 5 km N of Waramadan, Sipman & Aptroot 19243 (B); Mt. Latipu, Maas, Maas & Boyan 2608 (H); Region Cuyuni-Mazaruni, Chi-Chi Mountain range, Pipoly *et al*. 10220 (H); Region Potaro-Siparuni, Kaieteur Falls, DePriest *et al*. 9255 (H).

Notes: *Cladonia confusa* is in the Guianas most easily confused with shade forms of *Cladonia signata*. For differentiation see under that species. Among the species of subg. *Cladina*, *C. dendroides* and *C. rotundata* are very similar to *C. confusa* in morphology, but they differ by the presence of atranorin and the absence of usnic (mostly) and perlatolic acids.
The material from the Guianas deviates from Andean specimens of *C. confusa* because it forms very densely branched, wide heads (to about 5 cm wide) with thin branchlets, of a uniform grey-green colour; the branching is often largely dichotomous, instead of trichotomous, and apical browning is completely lacking. Similar forms occur throughout the Amazon basin, but not in the Brazilian Atlantic Forest, e.g., in the coastal lowlands of São Paulo, and may therefore deserve a special taxonomic status. They seem to intergrade with more typical populations (description see Ahti 2000) in Venezuelan Guayana, however.
Usnic acid-deficient plants (fo. *bicolor* (Müll. Arg.) Ahti) are common in most of the range. They lack the yellow tinge of the podetium colour, and they react negative with KC. So far they have not been found in the Guianas, and all available specimens belong therefore to fo. *confusa*.
The report of *Cladonia fallax* Abbayes (des Abbayes 1961: 117) most probably belongs here.

6. **Cladonia corallifera** (Kunze) Nyl., Flora 57: 70. 1874. – *Cenomyce corallifera* Kunze in Weigelt, Exs. (unnumbered, printed label with description). 1827-1828. –*Cladonia carnea* Hampe in Schomburgk, Reis. Br.-Guiana 1041. 1849, *nom. illeg.* – *Cladonia corallifera* var. *kunzeana* Vain., Acta Soc. Fauna Fl. Fenn. 4: 178. 1887, *nom. inval.* – *Cladonia corallifera* fo. *kunzeana* Vain., Acta Soc. Fauna Fl. Fenn. 14(1): 229. 1897, *nom. inval.* Type: Suriname, 1827, Weigelt s.n. (lectotype TUR-V 14165, designated by Ahti & Stenroos 1986, isolectotypes FH-Tuck, G, LE, MB, TNS, UPS, WRSL). – Fig. 10

Primary thallus persistent, often forming a characteristic, continuous crust on the substrate, consisting of small squamules up to 3 mm long, horizontal or appressed, laciniate, several times deeply divided into narrow, c. 0.2-0.3 mm wide lobes; upper side usually convex and white-maculate, lower side at the base usually orange to yellowbrown, connected to a yellow to orange hypothallus; terminal lobes sometimes fragile or transgrading into granules. Podetia up to 1.5 cm tall, 7-1.5 mm thick, of determinate growth, pale yellowish to greenish grey, forming scyphi; scyphi 2-7 mm wide, gradually dilating when short-stalked, or more suddenly when on longer stalks, occasionally proliferating from the margins; stalks up to three times as long as the scyphi. Podetial surface areolate-verruculose to granular, in its lowermost part sometimes verruculose with continuous cortex, towards the scyphus margin often abraded and revealing the pale brown stereome; squamules sometimes present at the base. Podetial wall 170-220 μm; cortex 0-10 μm; medulla 70-100 μm; stereome 80-170 μm;

Fig. 10. *Cladonia corallifera* (Sipman 39862 (B)). Bar = 2 cm.

central canal uneven, in part lacerate. Conidiomata frequent on margins of scyphi, subsessile, ovoid to dolioliform, 130-360 × 70-90 μm, slightly constricted at base, containing red slime. Hymenial discs frequent, up to 5 mm wide, red. Chemistry: usnic and thamnolic acids, with or without didymic acid (TLC of 14 specimens). Colour reactions: P+ yellow, K+ yellow, KC+ yellow.

Distribution and ecology: This is a neotropical endemic that occurs in the Amazonian lowlands in wide sense, from Venezuela and Colombia to Bolivia. It is especially common on white sands along the Amazon and its tributaries, as well as along the Orinoco. Widespread in the Guianas in open vegetation of all kinds, such as savanna, shrub savanna, man-made clearings, on poor, dry soil, preferably white sand, where it grows on decaying organic material such as rotten logs and termite nests, or directly on sand. In Guyana and Suriname it is reported from throughout the countries, from 10 to 2000 m alt. In French Guiana only known from the mountains of the interior, from 400 to 480 m alt. Over 200 collections studied (GU: 40; SU: 27; FG: 3).

Selected specimens: Exsicc.: Lichenotheca Latinoamericana 8, 59 (as *C. corallifera* var. *corallifera*). Guyana: S of Timehri, Maas & Westra 3532 (H, L); Mt. Roraima, Sipman & Aptroot 18897 (B). Suriname: Zanderij, Kramer & Hekking s.n. (H, L); Wilhelmina Gebergte, Irwin *et al.* 55523 (NY). French Guiana: Station des Nouragues, Bassin de l' Approuague-Arataye, face sud de l'inselberg, Roche Koutou - Bassin du Haut-Marouini, Hoff 5265 (B).

Notes: The bright red ascomata are frequently produced and together with the yellow-green thallus make this cup lichen a conspicuous and frequently collected species. It bears a superficial resemblance to and has been confused with *Cladonia coccifera* (L.) Willd., a species of cold environments, restricted in the Neotropics to high elevations.
The species has two sorediate relatives in the Guianas, *C. mollis* and *C. prancei*. The first is morphologically and chemically very similar to *C. corallifera*, and differs only by the presence of soredia on the podetia; the latter has narrow scyphi and some podetia may even be subulate, while usnic acid is often absent. *C. brasiliensis* (Nyl.) Vain. seems also closely related, and it differs by its narrow, long-stalked and much-branched scyphi, which lack both granules and soredia. Furthermore, the species seems restricted to eastern Amazonia and is not known from the Guianas.
Didymic acid is present only in specimens from Timehri. Outside the Guianas, didymic acid may be replaced by squamatic and/or barbatic acid (Ahti 2000).

33

Fig. 11. *Cladonia crassiuscula* (Duivenvoorden & Cleef 275 (B)). Bar = 2 cm.

Cladonia crassiuscula Ahti, Ann. Bot. Fenn. 23: 210. 1986. Type: Brazil, Pará, Serra do Cachimbo, Base Aérea do Cachimbo, 20 km N of the border with Mato Grosso on Cuiabá-Santarém highway, 430-480 m alt., Brako & Dibben 5853 (holotype INPA, isotypes B, H, NY, TNS, US, VEN). – Fig. 11

Primary thallus evanescent and mostly absent, consisting of short, crenulate squamules up to 2 mm wide. Podetia stoutish, 3-6 cm tall, of indeterminate growth, whitish to greenish-yellow, the extreme tips browned, emorient basal parts pale brown; growing in rather loose, spreading mats; richly branched; branching pattern very irregularly anisotomously, dichotomy or trichotomy, sometimes with isotomous tendencies; main stems not well differentiated, 1.5-3(-5) mm thick; branchlets short to long, thick, blunt, straight, divergent, spiny; adventitious branchlets common; axils perforated or closed, lateral holes and cracks frequent. Podetial surface matt, smooth to rugulose or ridged, continuous to areolate-variegate in necrotic parts. Podetial wall 240-300 μm, not very fragile; true cortex absent, but with 10-20 μm thick corticoid layer; medulla 220-280 μm, felty, replacing the proper stereome, which is lacking; surface of central canal matt, fibrose, striate and pulverulent. Conidiomata terminal, cylindrical, c. 400-500 × 100-200 μm, containing purple slime; conidia 8-10 × 1 μm, slightly arcuate. Hymenial discs rare, red-brown; mature spores not seen. Chemistry: usnic acid with thamnolic, barbatic, thamnolic + barbatic, or squamatic acids as major substances. Colour reactions: P+ yellow-orange, K+ yellow-orange, KC+ yellow or P-, K-, KC+ yellow.

Distribution and ecology: An Amazonian lowland species known from Venezuelan Guayana - where it reaches up to 1200 m alt. - and Pará, Brazil. It is especially abundant on the white sands with low to tall shrubs, along the upper Orinoco, growing more frequently in open glades rather than around shrubs. Not yet found in the Guianas. Over 20 collections studied.

Notes: In spite of considerable variation in size and habit, *C. crassiuscula* is easily distinguished from its sympatric allies, such as *C. spinea* and *C. steyermarkii*, by the thick, branchy podetia in which the stereome is replaced by a compact medulla (podetium wall soft). *C. sufflata* is similar, but almost unbranched.
A chemically analyzed (HPLC) specimen from Guyana was later reidentified as *C. spinea* (Ahti 2000: 324).

7. **Cladonia crustacea** Ahti, Fl. Neotrop. Monogr. 78: 191. 2000. Type: Brazil, Minas Gerais, Mun. Antônio Carlos, Instituto de São Miguel (5 km SE of Antônio Carlos railway sta.), 900-1000 m alt., Ahti, Krieger & Marcelli 48818 (holotype SP, isotypes H, US). – Fig. 12

Primary thallus persistent, forming large patches, appearing subcrustose, in part dissolving into white, granulose crust, otherwise consisting of tiny, densely crowded squamules, 0.5-1(-3) × 0.1-0.5 mm, strongly crenulate to featherlike, laciniate; upper side greyish white, lower side usually with orange or reddish brown vein-like streak; margins soon becoming abundantly granulose-sorediate and slightly incurved. Podetia rare and poorly developed, laminally on squamules, up to 3 mm long, 0.5 mm wide, of determinate growth, greenish grey, blunt, mainly esorediate. Podetial surface smooth, areolate-corticate. Podetial wall not studied. Conidiomata not seen. Hymenial discs not seen. Chemistry: thamnolic acid with unidentified substances (Ahti 2000). Colour reactions: P+ yellow, K+ yellow, KC-.

Distribution and ecology: Known so far from SE Brazil, where it grows on trunks and stumps in open forest from 10 to 2300 m alt. Here reported from French Guiana, on bark of trees in forest, from 150 to 200 m alt. About 10 collections studied (FG: 2).

Fig. 12. *Cladonia crustacea* (Aptroot 15576 (B)). Bar = 1 cm.

S e l e c t e d s p e c i m e n s : French Guiana: Saül, in and near the village, Aptroot 15193 (L); Montagne de Cacao, 45 km S of Cayenne, on big tree in dense forest, Aptroot 15576 (B, H).

N o t e s : The species is usually without podetia and distinctive in the field by forming conspicuous white crusts on trees. It is probably much overlooked. In Brazilian specimens the squamules may be up to 10 mm long.

8. **Cladonia dendroides** (Abbayes) Ahti, Ann. Bot. Soc. Zool.-Bot. Fenn. "Vanamo" 32(1): 29. 1961. – *Cladonia sandstedei* fo. *dendroides* Abbayes, Bull. Soc. Sci. Bretagne 16, Fasc. hors sér. 2: 99. 1939. – *Cladina dendroides* (Abbayes) Ahti, Beih. Nova Hedwigia 79: 38. 1984. Type: Brazil, Paraná, Jacareí ("Jacarehy"), Dusén 15231 in Malme, Lich. Austroamer. 226 (lectotype PC, designated by Ahti 1961, isolectotypes C, G, H, R, S, UPS, US). – Fig. 13

Primary thallus absent. Podetia 8-15 cm tall, of indeterminate growth, ashy grey, at extreme tips dark brown, forming irregular, 2-3 cm wide, densely branched, rounded heads, with mature internodes 2-4 mm long; branching type isotomous dichotomy, but basal parts often with distinct main stems, which can be up to 2 mm wide and angular; branch

Fig. 13. *Cladonia dendroides* (Pipoly et al. 9736 (B)). Bar = 2 cm.

axils open or closed; ultimate apical branchlets delicate, pointed; dead basal parts often (-always?) blackening. Podetial surface with thin felty ectal layer, ectal layer easily disintegrating, verruculose. Podetial wall c. 150 μm thick in main stems; stereome c. 80 μm; arachnoid hyphae 6-8 μm. Conidiomata containing red slime. Hymenial discs not observed, probably brown. Chemistry: atranorin and fumarprotocetraric acid, with traces of protocetraric acid, confumarprotocetraric acid (TLC of 9 specimens). Colour reactions: P+ fast red, K+ yellow, KC-.

Distribution and ecology: Known from coastal Brazil, from Paraíba to Santa Catarina, where it is rare. In the Guianas it was found in a few places on the lower sandstone tables of the Pakaraima Mts., the Rupununi savanna and the coastal white sand area near Linden, in Guyana. It grows on thin soil on flat rock surfaces or on sand in open places in scrub savanna, from 10 to 1000 m alt. Over 20 collections studied (GU: 15).

Selected specimens: Exsicc.: Lichenotheca Latinoamericana 108 (as *Cladina dendroides*). Guyana: Region Potari-Siparuni, Kaieteur Falls National Park, Ahti 52990 (BRG, H, US); Region Demerara-Mahaica, Linden Highway, Ahti 52928 (BRG, H, US); Region Demerara-Berbice, Linden-Soesdyke Highway, between Dora and Maibai Creek, Pipoly *et al.* 9736 (B, US).

Notes: *Cladonia dendroides* has been confused in the past with *C. sandstedei* Abbayes, which occurs in the Caribbean and Florida (Ahti 2000). It is closest to *C. rotundata*, with which it agrees in its isotomous, dichotomous branching, its chemistry and its thin outer felt layer. It differs by the presence of slightly developed main stems in its lower parts and of melanotic bases.
Difficult to distinguish may be *C. sprucei*, which has more distinct main stems and often reflexed ("combed") apical branches, and contains often psoromic acid as additional substance (traceable by TLC).
For differentiation from grey-coloured specimens of *Cladonia signata*, see note under that species.

9. **Cladonia densissima** (Ahti) Ahti & DePriest, Mycotaxon 78: 501. 2001. – *Cladina densissima* Ahti, Beih. Nova Hedwigia 79: 38. 1984. Type: Guyana, Pakaraima Mts., Mt. Aymatoi, 1150 m alt., Maas *et al.* 5702 (holotype U, isotype B, H). – Fig. 14

Primary thallus absent. Podetia 4-10 cm tall, of indeterminate growth, pale greenish-yellow, greyish-yellow or greyish-white, browned at the tips, extremely densely branched, with internodes 2-4 mm long when mature, forming 2-4 cm wide, cylindrical cushions; branching pattern anisotomous (in upper parts) to subisotomous dichotomy, rarely trichotomy; main stems usually distinguishable only in lower parts, 0.8-1.5 mm wide, occasionally somewhat flattened; apical and outer lateral branchlets attenuate and divergent or somewhat curly; axils closed or sometimes with a small hole; base not melanotic. Podetial surface verruculose throughout, the interspaces thinly arachnoid near the tips, mostly smooth towards the base. Podetial wall c. 100 μm thick, highly fragile, arachnoid surface layer (0-)10-20 μm, up to 60 μm including the algal layer; stereome 50-60 μm. Conidiomata containing red slime. Hymenial discs not observed, probably brown. Chemistry: usnic acid (rarely absent) and fumarprotocetraric acid, with traces of protocetraric and confumarprotocetraric acids, plus ursolic acid and accessory unknowns (TLC of 11 specimens). Colour reactions: P+ fast red, K- or + pale yellow, KC+ yellow (rarely KC-).

38

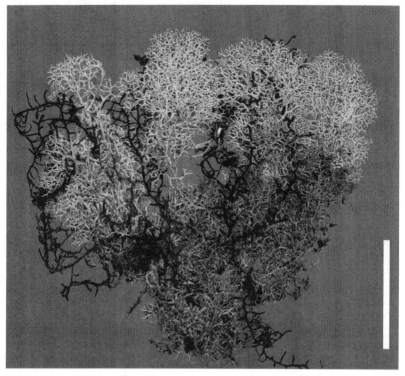

Fig. 14. *Cladonia densissima* (Sipman 39886 (B)). Bar = 2 cm.

Distribution and ecology: Mainly found in the Guayana Highland of Venezuela, at elevations from 120 to 2850 m alt., extending into adjacent Brazil and once found in Jamaica. At high elevations it occurs in open bogs, on rock outcrops and heathland, lower down in white-sand savannas. In the Guianas known from few localities on sandstone plateaux in the Pakaraima Mts. of Guyana and an outpost on the Tafelberg in Suriname, from 400 to 1200 m alt. Over 20 collections studied (GU: 20; SU: 1).

Selected specimens: Exsicc.: Lichenotheca Latinoamericana 5 (as *Cladina densissima*). Guyana: 10 km N of Waramadan, Sipman & Aptroot 19334 (B); Essequibo Basin, Kurupukari, A.C. Smith 2177 (S); Aymatoi, Maas, Mennega, ter Welle & Groen 5702 (H). Region Cuyuni-Mazaruni, Chi-Chi Mountain range, Pipoly 10202 (H). Suriname: Tafelberg, Maguire 24259L (BM, NY).

Notes: A rather variable species, with two fairly dissimilar, but not well separated morphs: some specimens contain very densely branched, columnar sprouts with semiglobose head and without clear main stems, resembling the well known boreal species *Cladonia stellaris* (Opiz) Pouzar & Vězda, while other specimens have more aggregated sprouts with few lateral branchlets, less pronounced semiglobose heads and prominent main stems, rather like *Cladonia ciliata* Stirt. An usnic acid-deficient strain, conspicuous in the field and recognized as fo. *decolorans* (Ahti) Ahti & DePriest (Ahti 1984), has been found in adjacent Venezuela. It is uncommon, but where present may form extensive clones. The presence of atranorin is reported from outside the Guianas. By its densely branched, globose heads with weakly developed main stems, *C. densissima* resembles *Cladonia spinea*. The latter differs by its corticate, completely smooth surface. Slender forms may resemble *C. confusa*, but can be recognized by the presence of fumarprotocetraric acid (P+ red).

10. **Cladonia didyma** (Fée) Vain., Acta Soc. Fauna Fl. Fenn. 4: 137. 1887. – *Scyphophorus didymus* Fée, Essai Crypt. Ecorc. cxviii, ci. 1825 ('1824'). Type: Dominican Republic, "Santo Domingo", Poiteau s.n. (lectotype G, designated by Ahti 1993, isolectotypes G, PC-Montagne, UPS). – Fig. 15

Cladonia muscigena Eschw. in Martius, Fl. Bras. Enum. Pl. 1(1): 262. 1833. Type: Brazil, Sellow (n. v.).
Cladonia vulcanica Zoll. & Moritzi, Natuur- Geneesk. Arch. Ned.-Indië 1: 396. 1844. – *Cladonia didyma* var. *vulcanica* (Zoll. & Moritzi) Vain., Acta Soc. Fauna Fl. Fenn. 4: 145. 1887. Type: Indonesia, Java, Banjoewangie, Zollinger 87 (lectotype L, designated by Stenroos 1986, isolectotype PC-Hue).
Cenomyce sphaerulifera Taylor, London J. Bot. 6: 185. 1847. Type: Guyana, Demerara, Parker s.n. (lectotype FH-Taylor, designated by Ahti 2000, isolectotypes FH-Dodge, G).

Primary thallus persistent, consisting of inconspicuous squamules, 1-2 × 1-1.5(-4) mm, split up irregularly into crenulate or lobulate laciniae, esorediate, upper side somewhat bluish green. Podetia (0.8-)1.5-2 cm long, 1-1.5 mm wide, of determinate growth, greenish to whitish or brownish, its basal, necrotic parts turning yellowish, simple or sparingly branched, mainly near the top; tips subulate to bluntish, not forming scyphi. Podetial surface ecorticate to slightly corticate at base, microsquamulose to partly granulose, esorediate to granulose-sorediate, usually largely denudated and smooth with exposed pale to dark brown, translucent stereome; microsquamules fragile, narrow, commonly projecting downwards or with incurved tips. Podetial wall 220-240 μm; cortex 0-12 μm; medulla 60-80 μm; stereome 120-160 μm; distinctly delimited; inner surface slightly grooved. Conidiomata on the primary squamules, rarely terminal

Fig. 15. *Cladonia didyma;* A. usual form (Samuels et al. 6136 (B)); B. fertile podetia with red hymenial discs (Sipman & Aptroot 19372 (B)); C. very slender podetia, common in the Guianas (Cremers 12366 (B)). Bar = 2 cm.

on the podetia, 100-250 μm wide, turbinate, short-stalked; containing red slime. Hymenial discs common, terminal, 0.5-4 mm wide, red. Chemistry: didymic acid with barbatic or thamnolic acids, occasionally didymic acid seemingly alone (TLC of 21 specimens). Colour reactions: P-, K-, KC- or P+ yellow, K+ yellow, KC-.

Distribution and ecology: A common pantropical species, extending into warm temperate areas in eastern N America, S America, E Asia, Australasia, and S Africa, growing preferably on decaying organic matter in open, often disturbed places with poor, acid soil. Widespread in the Guianas in similar habitats, from sea level to 2300 m alt. Over 1000 collections studied (GU: 23; SU: 9; FG: 6).

Selected specimens: Exsicc.: Lichenotheca Latinoamericana 9 (as *C. didyma* var. *vulcanica*). Guyana: N.W. Distr., Amakura R., de la Cruz 3527 (H, NY, US); Mt. Roraima, Sipman & Aptroot 18901 (B, BRG). Suriname: Zanderij, Florschütz & Florschütz 764 (L); Voltzberg, Florschütz 4576 (L). French Guiana: Savane Roche de Virginie, Bassin de l'Approuague, Cremers & Petronelli 11897 (B); Upper Marouini R.: 5 km WSW of Monpe Soula near three inselbergs, Samuels *et al.* 6136 (B, NY).

Notes: *Cladonia didyma* is one of the widespread, bacilliform, scarcely branched *Cladoniae* with persistent primary thallus, which come up often on disturbed places in nutrient-poor, sandy or peaty situations, together with the rather similar *C. subradiata*. It is easily recognized when it produces its red hymenial discs. In sterile state it is also very characteristic by its subulate podetia showing over most of their length the pellucid, brownish stereome. Chemistry also helps in its recognition: P+ yellow, K+ yellow (thamnolic acid strain) or P-, K- (barbatic acid strain). *C. subradiata* differs by its more cylindrical podetia, which remain more persistently granular to sorediose. Its hymenial discs, uncommonly produced, are brown. Finally, it reacts with P+ red, K- (fumarprotocetraric acid).
Reports of *Cladonia bacillaris* (Ach.) Genth and *C. macilenta* Hoffm. from the study area most probably belong here.

Cladonia flavocrispata Ahti & Sipman, Phytotaxa 93: 1. 2013. Type: Venezuela, Bolívar, Cerro Guaiquinima, near NE edge of upper plateau, 05° 54'N, 63° 27'W, c. 1250 m alt., rocky sandstone area with scrub on exposed ridge, H. Sipman 26772 (holotype VEN, isotypes B, H). – Fig. 16

Primary thallus evanescent, pale green, restricted to scattered, small squamules in the lower part of the podetia, c. 0.2-0.5 × 1 mm, simple or somewhat split up irregularly into crenulate laciniae, esorediate. Podetia

42

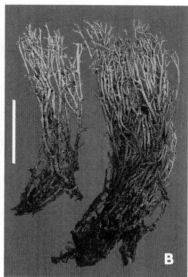

Fig. 16. *Cladonia flavocrispata;* A. type (Sipman 26772 (B)); B. Guyana specimen of uncertain affinity (Sipman 40299 (B)). Bar = 2 cm.

5-12 cm tall, of indeterminate growth, pale greenish grey, in lower parts strongly variegated, with alternating pale greenish grey and brown to black patches, usually forming cushions composed of dense, erect, 1.0-1.5 mm wide main stems; stereome soon brownish but not blackening inside; branching type anisotomous tricho- to polytomy; axils always perforated and only slightly swollen; branchlet tips short and darkbrown, usually under 0.5 mm long. Podetial surface matt, discontinuously very thinly corticate,with most of the stereome becoming bare with age, usually esquamose but some squamules may occur, particularly in fallen podetia; podetial squamules up to c. 1.5 mm wide, divided in c. 0.5 mm wide, elongate and crenulate lobes. Podetial wall 125-175(-200) μm thick, cortex 0-35 μm; medulla mostly absent; stereome 125-150 μm, pellucid, well-delimited. Conidiomata and hymenial discs not observed. Chemistry: thamnolic and usnic acids, sometimes with barbatic acids (TLC of 3 specimens). Colour reactions: P+ yellow, K+ yellow, KC+ yellow.

Distribution and ecology: The available samples suggest that this is a Guayana Highlands endemic, known so far with certainty only from Venezuela. It is found in humid sandstone tableland, and it grows on sandstone flats with open bog vegetation between ca. (400-)1000 and 2500 m alt. From the Guianas only three doubtful collections are known, from the sandstone plateau of the Kaieteur Falls (see notes). About 10 collections studied.

Specimens examined (paratypes): Venezuela: Amazonas, Depto. Atabapo, Cerro Marahuaca, Cumbre, 2480-2580 m alt., M. Guariglia *et al.* 1512 (H, VEN); Bolívar, Chimantá, Torono-tepui, 2100 m alt., Ahti *et al.* 45255 (H, VEN); Chimantá, 2130 m alt., Vareschi 9209 (H, VEN); Auyantepuí-Massif, Guayaraca, 1100 m alt., Vareschi & Foldats 6303 (H, VEN); Cerro Guaiquinima, 1000 m alt., Sipman 26514 (B, H, VEN); id., 1500 m alt., Sipman 27101, 27105 (B, H, VEN).

N o t e s : *Cladonia flavocrispata* is very similar to *C. hians* and could be considered its usnic acid-strain. However, it is also larger in size. Like *C. hians*, it belongs in section Perviae, as demonstrated by the regular perforations of the axils.
The species can be also easily confused with *C. vareschii*. The latter has a more intense yellow tinge and its cortex is somewhat thicker. Its apical branchlets stand at an obtuse angle (>90°) and bend away from each other immediately. A very reliable difference is the (often scarce) presence of squamules in *C. flavocrispata*.
Three Guianas specimens (Guyana: Potaro-Siparuni Region, Kaieteur Falls National Park, around the airstrip, c. 400 m alt., Sipman 40299, 40300, 40336 (B, BRG)) show a considerable resemblance, but deviate by the mostly closed axils not developing into funnels and the complete absence of squamules. In this repect they agree more with *C. vareschii* but lack the obtuse-angled apical branchlets and brownish colour. They may be more close to *C. spinea*, which lacks main stems, however.

11. **Cladonia furfuraceoides** Ahti & Sipman in Stenroos *et al.*, Cladistics 18: 243. 2002. Type: Guyana, Potaro-Siparuni Region, Kaieteur Falls National Park, near Kaieteur Guesthouse, 400 m alt., Ahti 53102 (holotype BRG holotype, isotypes B, H, NY, US). – Fig. 17

Primary thallus persistent or evanescent, consisting of incised, 1.5-3 mm wide squamules. Podetia 1-3 cm tall, 0.5-2 mm thick, of determinate growth, whitish-grey, hardly browning, not melanotic at base; cylindrical, unbranched or sparsely branched; tips usually forming narrow scyphi, sometimes subulate; scyphi c. 0.7(-1.5) mm wide, often dentate by marginal conidiomata, rarely proliferating. Podetial surface largely ecorticate but with small verruculae, which contain algal cells, rather loosely squamulose; squamules elongate, deflexed with incurved tips, with flat to convex upper surface, more or less dehiscent, c. 1.0-2.0 × 0.5 mm. Podetial wall 130-260 μm, rather soft; cortex present only on verruculae with algal cells, 20 μm thick; medulla 0-20(-60) μm; stereome 130-240 μm; central canal not striate. Conidiomata common, on scyphal margins, 200-380 μm long, 200 μm thick, constricted at base; containing hyaline jelly.

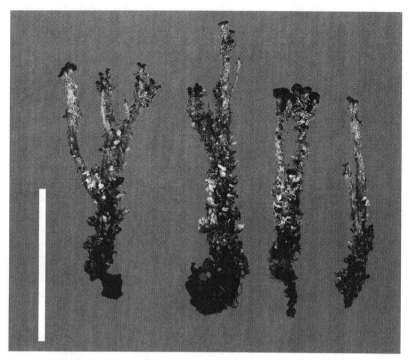

Fig. 17. *Cladonia furfuraceoides*; type (Ahti 53102 (B)). Bar = 2 cm.

Hymenial discs brown, 0.3-2.5 mm wide. Chemistry: fumarprotocetraric acid and traces of protocetraric and confumarprotocetraric acids, often also quaesitic acid, sometimes with distinctive fatty acids, once (Boom & Gopaul 7248) additionally atranorin (TLC of 16 specimens). Colour reactions: P+ red, K-, KC-.

Distribution and ecology: Outside the Guianas known from the Venezuelan part of the Guayana Highlands and adjacent Colombia and Brazil, reaching till near the Rio Negro. Mostly known from Guyana, from white sand areas and sandstone plateaux, but also on other mountains southward to the Kanuku Mts., with few scattered records from Suriname and French Guiana. Altitudinal range 30 to 500 m alt. It grows mainly on sandy, open places in scrub savanna, where it may be a colonizer of disturbed places. Over 50 collections studied (GU: 30; SU: 2; FG: 2).

Selected specimens: Guyana: Upper Mazaruni Distr., Kamarang, Boom & Gopaul 7248 (NY); Jawalla, Sipman & Aptroot 18351 (B); Region Potaro-Siparuni, Kaieteur Falls National Park, Ahti 53010, 53025 (H); Region Demerara-Mahaica, along Linden Highway, DePriest 9229

(H, US); Cuyuni-Mazaruni Region, 1 km W of Imbaidamai, Hoffman 1586 (H, US). Suriname: Nickerie district, area of Kabalebo Dam project, at km 96, Dakama savanna, Zielman 1475 (BBS, L). French Guiana: Roche Koutou - Bassin du Haut-Marouini, sommet de l'inselberg, de Granville *et al.* 9430 (B); Kourou, along Route Nat. 1 from Kourou to Sinnamary, around km 68, N of road after granite mine, Christenhusz *et al.* 2532 (B, TUR).

Notes: *Cladonia furfuraceoides* is most easily confused with *C. corymbites*, which is not with certainty known from the area (see under *C. cartilaginea*). It differs by the hard and thick, often denudated stereome, absence of small scyphi, and sometimes divaricate branching. It is probably more closely related to *C. polyscypha*, which it resembles much in shape and from which it differs by the production of well-developed squamules rather then soredioid granules on the podetia.
It has been united for a while (Ahti 2000) with *C. furfuracea* Vain., a common species in southern Brazil, and might be considered as its vicariant on the Roraima shield, northern Amazon. In typical forms, the differences are quite evident, but stunted plants might be diffcult to keep apart on morphological grounds (Stenroos *et al.* 2002: 243).
Typically, *C. furfuracea* forms taller and more slender podetia, from c. 0.4 × 1 cm when young to 20-40 × 0.8-1.2 mm when full-grown. The podetial squamules are distinctly smaller, usually up to about 0.5 mm long and 0.2-0.3 mm wide, and only occasionally exceed this size, up to about 1 mm length and 0.5 mm width. Small squamules may approach cylindrical isidia. The squamules tend to be much more densely placed. The scyphi often bear subulate extensions of the podetia, and only 1-3 hymenial discs; these may occasionally be found on subulate podetium tips. In contrast, the podetia of *C. furfuraceoides* frequently are only about 20 mm tall and 1-1.5 mm thick. The sparse podetial squamules reach normally 1 × 0.5 mm and may be up to 2 mm long. The scyphi may bear up to c. 5 hymenial discs. In addition, the primary thallus seems more persistent and is not granular.

Cladonia granulosa (Vain.) Ahti, Ann. Bot. Fenn. 23: 205. 1986. – *Cladonia subsquamosa* [var.] ß. *granulosa* Vain., Acta Soc. Fauna Fl. Fenn. 4: 448. 1887. Type: Colombia, Antioquia, Sonsón, 3000 m alt., Wallis s.n. (lectotype TUR-V 15142, designated by Ahti 1986, isolectotypes, G, PC-Hue).   – Fig. 18

Primary thallus usually persistent, consisting of narrow-lobed squamules up to 3 mm long, more or less granulose-sorediate at the margins. Podetia 2-3 cm tall, 0.5-2 mm thick, usually of determinate growth, light or ash-

46

Fig. 18. *Cladonia granulosa* (Sipman & Mandl 51346 (B)). Bar = 2 cm.

grey, usually browned for a large part; unbranched to usually 1-2 times successively branched, especially in the top parts; branching pattern anisotomous dichotomy or trichotomy, rarely tetrachotomy; axils mostly perforated, the youngest often closed, ascyphous or narrowly scyphoid, apical tips often acuminate. Podetial surface for the most part distinctly sorediate, granule size being mainly below 0.1 mm in uppermost part, lower down soredia immixed with microsquamules and towards the base macrosquamules present; corticate patches present near the base; ecorticate areas with distinct softer medullary layer on stereome. Podetial wall 320-400 μm thick, sorediate layer (including the algal glomerules) 80-120 μm; medulla 120-200 μm; stereome 120-180 μm, not distinctly delimited. Conidiomata common, 0.2 × 0.05-0.1 mm, containing purple slime. Hymenial discs rare, brown. Chemistry: thamnolic acid and commonly traces of decarboxythamnolic acid, rarely traces of barbatic acid (Ahti 2000). Colour reactions: P+ yellow, K+ yellow, KC-.

Distribution and ecology: Main distribution in the northern Andes from Venezuela to Peru, northwards extending to Costa Rica and Mexico, at altitudes between 1500 and 3100 m; scattered localities are reported on the coast of southern Chile. Not yet known from the Guianas, but with a few records from elevations above 2400 m alt. in adjacent Venezuela, from Cerro Marahuaca in Amazonas and from Auyántepui and Ptarí-tepui in Bolívar. Over 20 collections studied.

Note: This species is the only sorediate taxon from the section *Chasmariae* in the study area. The open axils characteristic for the section are frequently seen in the available material and will usually be sufficient to distinguish this species from superficially similar taxa from section *Cladonia*, e.g. the sorediate *C. polyscypha* and *C. subradiata*, and the squamulose *C. furfuracea*. In addition, the chemistry is distinct, thamnolic acid (P+ yellow, K+ yellow) against fumarprotocetraric acid (P+ red, K-). *C. subdelicatula* has the same chemistry, but it lacks open axils, and it is usually a smaller plant with thinner podetia.

12. **Cladonia guianensis** S. Stenroos, Ann. Bot. Fenn. 26: 255. 1989. Type: Venezuela, Bolívar, Dist. Piar, Macizo del Chimantá, valley between Torono-tepui and Chimantá-tepui, 2100 m alt., Ahti *et al.* 45262 (holotype VEN, isotypes H, NY, US).   – Fig. 19

Primary thallus persistent, consisting of 2-3 mm long, 1-3 mm wide and 0.4-0.5 mm thick squamules, sparsely divided into roundish lobes. Podetia well developed, borne laminally on primary squamules, 1-2.5 cm tall, 0.5-2 mm thick, of determinate growth, pale greenish brown, often

48

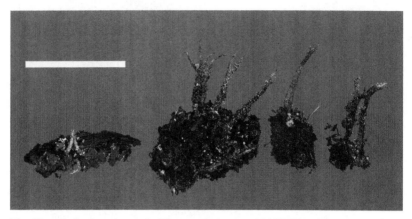

Fig. 19. *Cladonia guianensis* (Sipman & Aptroot 19139 (B)). Bar = 2 cm.

blackening in basal parts; subcylindrical; unbranched or apically 1-3 times branched when over 1 cm tall; sterile tips subulate. Podetial surface discontinuously corticate, squamulose, squamules up to 1.5 mm long, 0.2-0.3(-0.5) mm wide, white below, generally projecting downwards, seemingly dehiscent. Podetial wall 400-600 μm thick; cortex 30-150 μm; medulla 50-250 μm; stereome 100-400 μm, continuous but stranded, indistinctly delimited from the medulla, longitudinal strands filling the central cavity. Conidiomata 0.25-0.3 mm long, 0.2 mm wide, black, borne terminally or laterally on podetia, more rarely laminally on primary squamules; containing pale red slime. Hymenial disc fairly common, red, 1.5-2 mm wide. Chemistry: sekikaic and barbatic acids as major, constant compounds; accessory substances include 4'-0-methylnorsekikaic, 4-O-demethylbarbatic, 3-a-hydroxybarbatic, thamnolic and protocetraric acids, and atranorin; plus the hymenial pigment rhodocladonic acid; unidentified substances include the compounds M1, M4, M5, M6, M7, M19, M20 and M31 (Stenroos 1989a) (TLC of 1 specimen). Colour reactions: P-, K-, KC-.

Distribution and ecology: Apparently a Guayana Highland endemic, not rare in the corresponding parts of Venezuela and Brazil. Specimens were available from Parima, Huachamacari and Aratitiyope in Amazonas, from Chimantá, Churí-tepui and Guaiquinima in Bolívar, and from Serra Parima in Roraima. The species was found on acid humous soil over sandstone outcrops on treeless mountain tops, but also in montane forests, at 990-2400 m alt. In the Guianas, it is known only from Mt. Latipu in the Pakaraima Mts. at 1000 m alt. Over 10 collections studied (GU: 1).

Selected specimens: Guyana: Upper Mazaruni Distr., Mt. Latipu, c. 8 km N of Kamarang, Sipman & Aptroot 19139 (B, BRG).

Notes: This species is characterized by its sturdy, squamulose podetia, with squamules pointing downwards, and its central cavity filled with stereome strands. The variable chemistry and thick cortex indicate that it belongs to the *C. miniata* group. Within the group it stands apart by its poorly developed primary thallus and large podetia.

Cladonia hians Ahti, Fl. Neotrop. Monogr. 78: 284. 2000. Type: Venezuela, Bolívar, Macizo del Chimantá, Acopán-tepui, 1950 m alt., Ahti *et al.* 45067 (holotype VEN, isotypes B, DUKE, FH, H, MERF, NY, US, VEN). – Fig. 20

Primary thallus evanescent, pale green, restricted to scattered, small squamules in the lower part of the podetia, c. 0.2-0.5 × 1 mm; simple or somewhat split up irregularly into crenulate laciniae; esorediate. Podetia 3-9 cm tall, 0.3-1.2 mm wide, of indeterminate growth, glaucous, whitish or brownish-grey, in older parts usually strongly variegated, with alternating grey and brown to black patches, near the base sometimes with cyanescent spots; stereome soon brownish but not blackening inside; branching type anisotomous polytomy or less frequently isotomous dichotomy; axils always perforated (often gaping), sometimes producing scyphoid structures 0.3-1.5(-3) mm wide, with 3-6(-9) marginal teeth or longer proliferations. Podetial surface matt, discontinuously very thinly corticate, with most of the stereome becoming bare with age, usually esquamulose but small, narrowly laciniate squamules may occur even near tops of podetia. Podetial wall 125-175(-200) μm thick, cortex 0-35 μm; medulla mostly absent; stereome 125-150 μm, pellucid, well-delimited. Conidiomata common, cylindrical, 200-400 × 30-100 μm; containing hyaline slime. Hymenial discs at ends of thickened podetia, light brown, 0.5-1 mm wide. Chemistry: thamnolic acid and sometimes traces of barbatic and 4-0-demethylbarbatic acids (Ahti 2000). Colour reactions: P+ yellow, K+ yellow, KC-.

Distribution and ecology: Endemic to the Guiana Shield and surroundings, with records from Colombia (Araracuara) and Venezuela (Bolívar and Amazonas), where it was found mainly on sandstone plateaux at low and mid elevations to c. 2000 m alt. Not yet known from the Guianas. About 10 collections studied.

50

Fig. 20. *Cladonia hians* (Sipman 27136 (B)). Bar = 2 cm.

Fig. 21. *Cladonia huberi;* isotype (Ahti et al 45165 (B)). Bar = 2 cm.

Notes: *Cladonia hians* has a poorly developed cortex and medulla, which makes that the stereome becomes easily exposed, and this causes the podetia to become variegated when the stereome is brownish. It may also produce numerous regular scyphoid funnels, particularly in well-illuminated high-altitude habitats, though in shade and at lower elevation the funnels are narrower and less regularly developed. In general habit it is rather similar to forms of the temperate *C. crispata*, a related species, which lacks the variegated pattern on the podetia.

Among the Guianas Cladoniae, *C. sipmanii* can be very similar but most of its higher axils are closed. More significantly, its cortex and medulla are more strongly developed, often as thick as the stereome, and present over the whole podetum length, so that no variegate pattern develops even near the base of the podetia. In *C. hians* cortex and medulla are often largely absent, except in the upper part of the podetia, and in the algal cell clusters, thus causing a variegated aspect of the podetia. The extremes are quite distinct morphologically, and also ecologically: strongly variegate and funnel-bearing plants occur in mossy bog at high elevation, while typical *C. sipmanii* occurs on bare sand of white sand savannas in the coastal lowlands. However, intermediate plants occur, with thin cortex and medulla, but without funnels and with more or less variegated colouring, which may be difficult to classify. These seem to be, at least in part, young individuals, where the characteristic form of the podetia has not yet developed. *C. huberi* is probably another close relative, see below.

Cladonia huberi Ahti, Fl. Neotrop. Monogr. 78: 287. 2000. Type: Venezuela, Bolívar, Macizo del Chimantá, NE section of Acopán-tepui, 1950 m alt., Ahti, Huber & Pipoly 45165 (holotype VEN, isotypes B, H, NY, US). – Fig. 21

Primary thallus unknown. Podetia 8-15 cm tall, in main stems 0.3-0.8 mm thick, of indeterminate growth, whitish to brownish-grey, easily browned in exposed top parts but never predominantly blackish, at base somewhat melanotic, forming erect cushions but not very distinct rounded heads; axils mostly narrowly perforated; branching type isotomous or anisotomous dichotomy, occasionally anisotomous tricho- or tetrachotomy; ultimate lateral branchlets somewhat curly, tapering to very fine, black tips (0.5-)1-2 mm long, 0.1-0.2 mm thick. Podetial surface matt, appearing smoothly corticoid, with some verruculae and maculae. Podetial wall 100-150 μm; corticoid layer 20 μm; stereome 80-100 μm, distinct but not strong; inner surface papillulate. Conidiomata not seen. Hymenial discs not seen. Chemistry: thamnolic and barbatic acids and probably traces of squamatic acid (Ahti 2000). Colour reactions: P+ yellow, K+ yellow, KC-.

Distribution and ecology: An infrequent endemic from the Guayana Highlands, known only from Chimantá and Aprada-tepui in Bolívar, Venezuela, where it was found growing in bog at 2000-2500 m alt. Not yet found in the Guianas. Three collections studied.

Notes: *Cladonia huberi* is fairly similar to *C. pulviniformis*. Apart from its completely conglutinated, albeit weak, stereome, *C. huberi* differs because it does not form clear semiglobose heads and develops more pronounced main stems. Its closest relatives may be *C. hians* and *C. sipmanii*. The first is usually easily distinguished by its funnels and variegated podetia, the latter has usually a thicker medulla and lacks the fine, blackish branchlet tips.

The dot inside Guyana on the map of Ahti (2000: fig. 169) is misplaced; the actual specimen is from Venezuela.

13. **Cladonia hypoxantha** Tuck., Proc. Amer. Acad. Arts Sci. 5: 393. 1862. – *Cladonia coccifera* subsp. *hypoxantha* (Tuck.) Vain., Acta Soc. Fauna Fl. Fenn. 4: 174. 1887. Type: Cuba, Santiago de Cuba: Monte Verde, Wright, Lich. Ins. Cubae 41 (lectotype FH-Tuck, designated by Ahti 1993, isolectotypes BM, M, PC, UPS, US).

– Fig. 22.

Fig. 22. *Cladonia hypoxantha* (Ahti 53352 (B)). Bar = 0.5 cm.

Primary thallus persistent, forming depressed mats, consisting of elongate squamules up to 5 mm long, finely lacinulate and crenulate, with laciniae 0.5-1 mm wide, flat, convex to cylindrical towards the base, esorediate or sorediate below, lower side usually with vein-like ochraceous stripe. Podetia not seen, acc. to Ahti (2000) very short, 0.3-0.7 (-4) cm in height, 1 mm thick, of determinate growth, ascyphose but more often truncate than subulate, or scyphose, with scyphi 0.5-3 mm wide. Podetial surface continuously corticate, but with minute, slightly elevated areolae separated by narrow grooves, esorediate to sparsely granulose-sorediate, rarely squamulose. Hymenial disks red. Chemistry (after Ahti 2000): thamnolic, tr. decarboxythamnolic acids, bellidiflorin (accessory), rhodocladonic acid (in hymenial disc). Color reactions: P+ yellow, K+ yellow, KC-.

Distribution and ecology: Known so far from Florida and the larger Antillean islands, Cuba, Hispaniola and Trinidad. Found once in Guyana on charred wood, at c. 700 m alt. An inconspicuous species that is possibly much overlooked. It grows on trunks and logs. About 20 collections studied (GU: 1)

Selected specimen: Guyana: Region Potaro-Siparuni, 0.5-2.0 km NW of Paramakatoi, Karibon Creel Trail, Ahti *et al.* 53352 (B, BRG, H, US).

Notes: Most easily confused with juvenile thalli of *C. corallifera*, which share the elongate squamules with ochraceous basal pigment and presence of thamnolic acid. In *C. corallifera* the primary thallus is less dissected and never sorediate, it soon produces cup-shaped podetia and it grows mainly on sandy soil.
A very similar species is *C. crustacea*, which shares the scarcety of podetia, the presence of soredia on the squamules and of thamnolic acid, and the corticolous habitat. It differs in the strongly applicated, not lacinulate squamules without ochraceous ventral stripes.

14. **Cladonia isidiifera** Ahti & Sipman, Phytotaxa 93: 1. 2013. Type: Guyana, Upper Mazaruni Distr., E-bank of Waruma R., 20 km S of confluence with Kako R., in c. 20 m . tall, virgin, riverine forest, on overhanging tree along river, c. 550 m alt., 11 Feb. 1985, H. Sipman & A. Aptroot 18671 (holotype B). – Fig. 23

Fig. 23. *Cladonia isidiifera;* type (Sipman & Aptroot 18671(B)). Bar = 1 cm.

Primary thallus persistent, consisting of 1-4 mm long, horizontal to ascending squamules which are deeply divided in c. 0.5-1.0 mm wide and c. 0.3 mm thick, slightly crenulate lobes, with flat to concave, green upper side and white or pale brown, shiny lower side, with pale hypothallus, often isidiate along the tips of erect, up to 5 mm long and 2 mm wide extensions of the lobes; isidia short-cylindrical, c. 0.3 mm long and 0.15 mm wide, glossy. Podetia, conidiomata and hymenial discs not known. Chemistry: barbatic and 4-O-demethylbarbatic (minor) acids. Colour reactions: P-, K-, KC-.

Distribution and ecology: Known only from a single specimen found in the Upper Mazaruni Distr., Guyana, on an overhanging trunk along a stream at c. 550 m alt. in mossy forest. (GU: 1).

Specimen examined: type only.

Notes: The presence of cylindrical isidia is so unusual in the genus *Cladonia*, that there is no doubt that the only available specimen belongs to an undescribed species. The thallus squamules show that it is related to *C. miniata*. Morphologically it bears most similarity with *C. ahtii*. This species is sorediate instead of isidiate, and it lacks the peculiar lobe extensions. In the *C. miniata*-group there is one more isidiate species, *C. caribaea* S. Stenroos. This has coralloid to flattened, not cylindrical isidia.

15. **Cladonia maasii** Ahti & Sipman, Phytotaxa 93: 1. 2013. Type: Suriname, 2 km N of Kamisa Falls in Nickerie R., shrub savanna, 2 July 1968, P.J.M. Maas 3378 (holotype B, isotype H, L).
– Fig. 24

Primary thallus squamulose, evanescent, squamules c. 0.1 mm diam. Podetia 4-6(-10) cm tall, 0.3-1 mm thick, rarely with scatted, small squamules, of indeterminate growth, basic colour whitish grey but exposed parts may become dark brown, forming erect, dense, flat, interwoven mats; branching type anisotomous dichotomy, to a lesser degree trichotomy or tetrachotomy; main stems in part distinguishable; axils commonly perforated; tips erect, divaricate, acute. Podetial surface clearly corticate; cortex smooth to somewhat rugulose, dull, with some ecorticate patches; soredia and squamules lacking. Podetial wall thin, c. 100-200 μm; cortex 50 μm; medulla 50-100 μm; stereome c. 100 μm. Hymenial discs not seen. Conidiomata scarce, at the end of podetia, 200-300 × 100-150 μm, ovoid to ampullaceous, constricted at base, black, content not seen. Chemistry: fumarprotocetraric, tr. protocetraric, homosekikaic and/or sekikaic acids. Colour reactions: P+ red, K-, KC-.

Distribution and ecology: A Guianan endemic, as far as presently known, represented only by two collections from Suriname and one from Guyana, in shrub savanna, from c. 200 to 1000 m alt. Three collections studied (GU: 1; SU: 2).

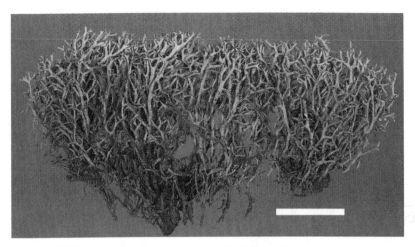

Fig. 24. *Cladonia maasii;* type (Maas 3378 (B)). Bar = 2 cm.

Specimens examined (paratypes): Guyana: Ituru Yawaruki savanna, white sand areas, Abraham 123 (BM) (TLC (F. Oberli, G): fumarpr., protoc., homosekikaic, sekikaic). Suriname: Natuurreservaat Brinckheuvel, Sabanpasi savanne complex, Teunissen & Wildschut LBB 11403 (L).

Notes: *Cladonia maasii* is named in honour of the collector, Prof. Dr. Paul J.M. Maas, who discovered several important *Cladonia* sites in the Guianas.
The lichen resembles *C. peltastica* but contains fumarprotocetraric acid and tends to get slightly brown and has no usnic acid. Its branchlets are also more robust, with perforated axils, and the ramification is less dense than in *C. peltastica*. A specimen from Guyana (Mt. Latipu near Kamarang, Sipman & Aptroot 19164) is included here with doubt. It agrees in chemistry but the podetia are more slender.

16. **Cladonia meridionalis** Vain. in Zahlbruckner, in Schiffner, Denkschr. Kaiserl. Akad. Wiss., Wien, Math.-Naturwiss. Kl. 83: 136. 1909. Type: Brazil, São Paulo, Raiz da Serra, 1901, Schiffner 9 (lectotype TUR-V 14175, designated by Ahti 1993, photo US, isolectotypes BM, W, WU).   – Fig. 25

Primary thallus persistent, consisting of large, lobed squamules, 15-25 × 2-8 mm, easily incurving, becoming subtubular, often clearly stalked, veiny, sorediate on lower side along the tip margins, with basal

58

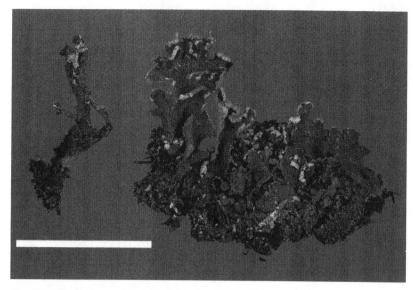

Fig. 25. *Cladonia meridionalis* (Sipman 40302 (B)). Bar = 2 cm.

stalks often 1 mm thick, cartilaginous, sometimes hollow, podetioid, implanted laminally; soredia farinose to granular, occasionally scarce. Podetia sparsely produced, often phyllopodiate, 1-4 cm tall, 0.7-2.5 mm thick, of determinate growth, glaucescent-grey, unbranched to slightly branched, especially near tips, not forming scyphi, sometimes tipped with squamules. Podetial surface almost continuously corticate, with some sorediate patches; slightly squamulose. Podetial wall anatomy not studied. Conidiomata on primary squamules or at tips of podetia, sometimes clearly stalked, cylindrical to ovoid, 120 × 100 μm in size; blackish-brown to red at ostiolum. Hymenial discs uncommon, red, 1-2 mm wide. Chemistry: obtusatic acid as major compound, with barbatic acid and trace amounts of unidentified compounds probably from the sekikaic acid aggregate (TLC of 4 specimens). Colour reactions: P-, K-, KC-.

Distribution and ecology: Outside the Guianas only known from the states of São Paulo and Paraná in SE Brazil, and probably much overlooked. Widespread but scattered in the interior of the Guianas. It grows on rotten wood, termite mounds (especially in Guyana) or earth banks in shade in open woodland, from c. 200 to 800 m alt. About 15 collections studied (GU: 7; FG: 2).

Selected specimens: Guyana: Region Potaro-Siparuni, Kaieteur Falls National Park, DePriest 9317 (BRG, H, US); Upper Mazaruni Distr., trail from Kamarang R.i to Pwipwi Mt., c. 10 km N of Waramadan, Sipman & Aptroot 19467 (B, BRG). French Guiana: Saül, 2 km S of the village, sentier Limonade, Montfoort & Ek 1476 (L); Inselberg de Montagne de la Trinité, sommet NE, de Granville *et al.* 6251 (B).

Notes: *Cladonia meridionalis* is an unmistakable species because of its unusually large, subtubular, sorediate primary thallus squamules. The unusual stalks of the squamules are distinct only in well developed specimens. Podetia and red hymenial discs are sparingly produced, and the latter were not yet found in collections from the Guianas. It tends to grow in small individuals with only primary thallus, and is therefore probably easily overlooked. That might explain its seemingly disjunct occurrence in SE Brazil and the Guianas, not a usual distribution pattern in phytogeography.

17. **Cladonia miniata** G. Mey., Nebenst. Beschaeft. Pflanzenk. 149. 1825. Type: destroyed in B. Neotype: Brazil, Rio de Janeiro, Mun. Itatiaia, Parque Nac. Itatiaia, 5 km ENE of Alto da Serra (12 km along road BR 485), vic. Hotel Alsene, 2320-2350 m alt., Ahti & Windisch 45975 (holoneotype SP, designated by Stenroos 1989a, isoneotypes H, NY). – Fig. 26

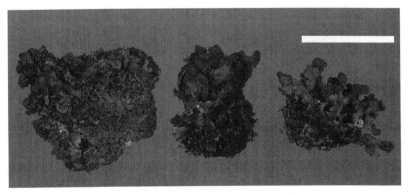

Fig. 26. *Cladonia miniata* (Ahti et al. 45064 (B)). Bar = 2 cm.

Primary thallus persistent, consisting of sparsely lobed, 4-7 mm long, 2.5-4 mm wide squamules, which are unusually thick (0.4-0.6 mm) and not showing conspicuous hygroscopic movements when wetted; upper surface matt, olive green to brown, verruculose; lower surface bright orange, often with distinct, longitudinal veins, sometimes with a loose, white net of hyphae at the very surface; lobe margins often sparsely granulose; cortex 150-250 μm thick, translucent, hard-cartilaginous, with verruculae 100-200 μm high; algal layer distinct, 30-50 μm; medulla orange, 50-380 μm (including a 20-30 μm thick white layer below the algae). Podetia poorly differentiated from the squamules, mostly flattened, dorsiventral, up to 3 cm tall, 2 cm wide, of determinate growth, olive green to brown, usually abruptly flaring, up to 2 cm wide, dorsiventral, flattened, with web-shaped structures, borne marginally or more rarely laminally on the primary squamules, which may elongate and roll up to form the tube of a podetial stalk; the central canal of the podetium may remain open on the ventral side of the squamules. Podetial surface continuously corticate, rugulose, often cracked. Podetial wall very thick, 470-900 μm, cortex 150-350 μm, algal layer 20-100 μm, medulla (100-)300-600 μm (in some parts even thicker, including a 10-20 μm thick white layer below the algae), orange; hyphae often loosely filling the inner podetial cavity; cylindrical stereome absent or poorly developed, skeletal tissue represented by internal, longitudinal strands, which are partly surrounded by medullary hyphae. Conidiomata dolioliform, 200-250 × 150-200 μm, black to red, borne marginally or laminally on the primary squamules, or rarely on the podetia; containing red slime. Hymenial discs commonly produced, red, forming up to 7 mm wide agglomerations; primordia borne on the primary squamules. Chemistry: barbatic acid (TLC of 1 specimen). Colour reactions: P-, K-, KC-.

Distribution and ecology: Restricted to the Neotropics, occurring in the Brazilian Highlands, the northern Andes, the Guayana Highlands, and perhaps Jamaica, at elevations of c. 1000-2000 m alt. in SE Brazil and 1800-3000 m alt. in the Andes. Venezuelan specimens were available from the cerros Duida and Marahuaca in Amazonas and from Chimantá, Auyan-tepui and Guanay in Bolívar, between 1200 and 2500 m alt. It grows mostly on rotting wood, but also on trunks of living trees up to 2 m from the ground, on thin soil over rocks and, rarely, on sandy soil. In the Guianas, it is so far known from a single specimen, from Mt. Latipu in the Pakaraima Mts., at 1000 m alt. Over 100 collections studied (GU: 1).

Selected specimen: Guyana: Upper Mazaruni Distr., Mt. Latipu, c. 8 km N of Kamarang, Sipman & Aptroot 19061 (B, BRG).

Notes: The orange medulla of the squamules makes this lichen unmistakable. Its closest relatives in the Guianas, *C. meridionalis* and *C. secundana*, have a white medulla. The orange medulla makes the species also very conspicuous, so that it is easily noticed also by non-specialists. Thus the low number of available specimens is a reliable indication that the species is rare in the Guianas.

The secondary chemistry is highly variable (see Stenroos 1989a), but rhodocladonic acid (the orange pigment) is a constant medullary substance and barbatic acid a subconstant major compound; didymic acid is a rare major substance; accessory minor compounds include rather commonly 4-0-demethylbarbatic and 3-a-hydroxybarbatic acids, more rarely squamatic, baeomycesic, thamnolic, obtusatic, fumarprotocetraric, protocetraric, sekikaic, 4'-O-methylnorsekikaic, condidymic, and subdidymic acids, as well as 9 unidentified compounds. This is the most variable secondary (primarily phenolic) chemistry reported for any *Cladonia*, and is unusual among lichens in general.

18. **Cladonia mollis** Ahti & Sipman, Phytotaxa 93: 1. 2013. Type: Guyana, Demerara-Mahaica Region, on Linden Highway, km 7 from Soesdyke, by end of trail to Marudi Creek Resort, 06° 31'N, 58° 12'W, 10 m alt., on burnt stump in secondary woodland with savanna patches on white sand, 1996, T. Ahti, R. Lücking & H. Sipman 52910 (holotype BRG, isotypes B, H, US, VEN). – Fig. 27

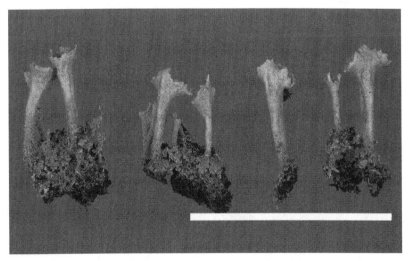

Fig. 27. *Cladonia mollis;* isotype (Ahti et al. 52910 (B)). Bar = 2 cm.

Primary thallus squamulose, consisting of green, flattish squamules with convex lobes and necrotic bases turning orange. Podetia 1-2(-3) cm tall, of determinate growth, whitish to greenish yellow, always forming scyphi; scyphi 1-4 mm wide, usually single but with age proliferating from the margins and sometimes forming a second scyphus at end of the proliferations, just below the short-stalked apothecia. Podetial surface totally ecorticate or little corticate at the very base, otherwise very rough due to coarse granules and often very densely beset with microsquamules, all of which dissolve into a thick layer of loose, finely granulose soredia towards the tops. Podetial wall not measured. Conidiomata formed on margins of young scyphi, typicall black, cylindrical, shortly stipitate; with purple slime inside. Hymenial discs unusual, at margins of scyphi forming 1-2 mm wide purple disks. Chemistry: K+ yellow, PD+ yellow, containing usnic and thamnolic acids, as well as the purple pigment rhodocladonic acid in hymenial discs and conidiomata.

Distribution and ecology: *Cladonia mollis* is known from the Guianas and northern Brazil, but is expected to be more widespread in Amazonia. In Guyana it was observed on rotten wood and white sand in forest clearings and savannas, from 10 to 500 m alt. Over 25 collections studied (GU: 11; SU: 2; FG: 1).

Selected specimens (paratypes): Guyana: Demerara-Mahaica Region: type locality, Ahti 52900 (B, BRG, H, US), 52912 (BRG, H, US); Cuyuni-Mazaruni Region, Bartica, Linder 810 (FH, TUR-V 14099, W, as "*Cladonia Guiana* Vain." in herb.); Potaro-Siparuni Region, Kaieteur Falls National Park, DePriest 9316 (BRG, H, US); Tukeit, DePriest 9361 (BRG, H, US); Basin of Essequibo R., Kurupukari, A.C. Smith 2178 (NY); Upper Demerara-Berbice Region, Mabura Hill, along logging road WS 1200, Waraputa compartment, DePriest 9179 (BRG, H, US); Upper Mazaruni Region, Makwaima savanna near Mayaropai, at Kako R., wet savanna on white sand, Sipman & Aptroot 18538 (H, L). French Guiana: Inselberg de Montagne de la Trinité, NE summit, on dead wood in rock savanna, de Granville *et al.* 6252 (B, L). Suriname: No locality, [1827-28] C. Weigelt in Herb. Schweinitz (PH); Sabanpasi savanna complex, Nature Reserve Brinckheuvel, summit of Brinckheuvel, on coarse, white sand, Teunissen & Wildschut LBB 11933 (H, L); Jodensavanne, Benjamins (L), Stahel (L, REN).
Brazil: Amazonas: Manaus-Itacoatiara road km 18, under 100 m alt., in secondary campina forest, Richards 6949 (BM); Pará: Serra do Cachimbo, Base Aérea do Cachimbo, c. 20 km N of the border with Mato Grosso on Cuiabá-Santarém highway (RB-163), c. 09° 22'S, 54° 54'W, broad, sandy, level riverine plain, Brako & Dibben 5810 (H, INPA, NY).

Notes: The species is distinguished from the very similar *C. corallifera* by the production of distinct soredia almost throughout the surface of podetia. However, it may be difficult to distinguish from granulose morphs of *C. corallifera*. In Guiana we encountered the two species growing together in some places, where they appeared to be distinct. Also E. A. Vainio recognized the species under an unpublished herbarium name (see above). All the specimens of *C. mollis* that were chemically studied contained thamnolic acid, while in *C. corallifera* the chemistry is more variable (Ahti 2000).

C. mollis is also closely related to *C. prancei*, another sorediate derivative of *C. corallifera*, which in addition has podetia forming narrow scyphi, sometimes becoming subulate, with a low content of usnic acid often giving the podetia a pale grey colour without yellowish tinge. Sorediate squamules have been found only in the specimen Sipman & Aptroot 18538.

19. **Cladonia peltastica** (Nyl.) Müll. Arg., Flora 63: 260. 1880. – *Cladina peltastica* Nyl., Flora 57: 70. 1874. Type: Brazil, Amazonas, Umirisál opposite to Manaus at mouth of Rio Negro, 1868, Spruce, Lich. Amaz. And. 22 (lectotype H-NYL 37612, designated by Huovinen & Ahti 1986, isolectotypes BM, G, NY, W). – Fig. 28

Primary thallus evanescent, rarely visible, consisting of tiny squamules 1-4 × 0.5 mm. Podetia 1-10 cm tall, 0.2-0.4(-0.6) mm thick, when fertile longer and swollen, up to 1.5 mm thick, of indeterminate growth, greenish-yellow to rarely ash-grey, at extreme tips browned, not melanotic at base, richly branched, branching type mainly anisotomous or subisotomous dichotomy, forming extensive, dense, flat to somewhat convex colonies; axils closed or rarely perforated; branchlets stiff, tending to grow straightly upright; tips subulate, not forming scyphi. Podetial surface smoothly corticate, shiny, occasionally in part verruculose-areolate; esquamose or with scattered squamules which may become deeply divided into linear, 0.2-0.3 mm wide slips. Podetial wall 100-170 μm, cortex 20-40 μm, medulla 40-70 μm, stereome 40-60 μm; surface of central canal smooth. Conidiomata at tips of podetia, often solitary, 130-160 μm thick, cylindrical, hardly constricted at base; containing red (or hyaline?) slime. Hymenial discs infrequent, in groups at podetial tips but rarely aggregate, 0.3-0.5 mm wide, peltate, dark to pale brown. Chemistry: an unusually large complex of chemotypes, consisting of at least 16 different combinations of one or more of the following substances: usnic, thamnolic, barbatic, squamatic, homosekikaic, boninic, didymic acids. The substances may be accompanied by some of their usual satellites, and some belong to complexes where the individual substances are less easy to identify (TLC of 67 specimens). Further chemotypes may be present in the Guianas as the material is not completely investigated. Colour reactions: P+ yellow or P-, K+ yellow or K-, KC+ yellow or KC-.

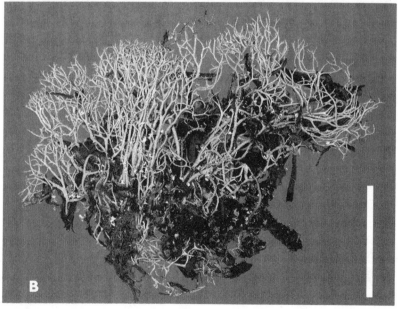

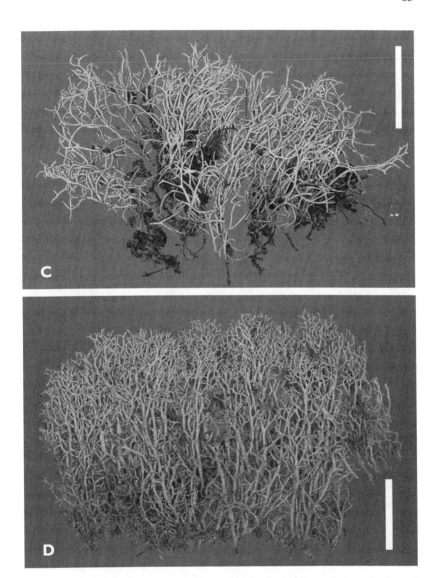

Fig. 28. *Cladonia peltastica;* A. thallus with slender sterile podetia and enlarged fertile podetia (Sipman & Aptroot 19149 (B)); B. shade form with abundant squamules (Sipman & Aptroot 18539 (B)); C. usnic acid-containing thallus from clearing (Maas et al. 7583 (L)); D. thallus with short internodia (Jansen-Jacobs et al. 5420 (B)). Bar = 2 cm.

Distribution and ecology: Restricted to the Neotropics, and occurring mainly in the Amazonian lowlands (up to 1000 m) in Colombia, Venezuela, Brazil, Peru and Bolivia. Widespread and common terrestrial lichen throughout Guyana and Suriname, so far not reported from French Guiana, growing often in abundance on slightly shady, mossy spots in open woodland or on poor, usually sandy soil in scrub savanna, from 10 to 1000 m alt. Over 200 collections studied (GU: 63; SU: 16).

Selected specimens (roman numbers refer to chemical combinations in Table 1): Exsicc.: Lichenotheca Latinoamericana 112. Guyana: Yapacooma, Bartlett 8041 (BRG); Plantation Vryheid, Linder 967a (FH, TUR-V 15282), IV; Kurupukari, Smith 2177 (NY), XIII; Upper Mazaruni Distr., Boom & Gopaul 7665 (NY), XII; Kaieteur Falls, DePriest et al. 9270 (H, US); Kamarang R. - Pwi Pwi Mt., Sipman & Aptroot 19321 (B), V; Barima R., de la Cruz 3415 (NY, US), V, VII; Mora Landing, de la Cruz 1836 (BM, US), XII; Mt. Latipu-foot, Maas & Westra 4173 (H), XV; Region Cuyuni-Mazaruni, Chi-Chi Mountain range, Pipoly 10209 (H); Region East Demerara, Yarowkabra Housing scheme, Pipoly 8465 (H); Region Demerara-Mahaica, Linden Highway, DePriest 9223 (H); Demerara-Mahaica Region, Yarowkabra, Pipoly 11788 (H, US). Suriname: Jodensavanne, Wenck s.n. (M, S, UPS), VII; Cassiporakreek, Benjamins s.n. (L); Zanderij, Florschütz & Florschütz 766 (L), IX; Brinckheuvel, Teunissen & Wildschut LBB 11932 (L); Wilhelmina Mts., Stahel BW 7043 (H); Brownsweg, Lanjouw 168 (H), VII.

Notes: *Cladonia peltastica* is a highly variable species both in morphology and chemistry. Attempts to distinguish parts of it as distinct species have failed so far, although some forms were separated for a while as *Cladonia submedusina* Müll. Arg. (*Cladonia medusina* (Bory) Nyl. var. *submedusina* (Müll. Arg.) Vain. ex Zahlbr.). Those form often large, dense mats of finely branched podetia with upward tips, sometimes mixed with a few emergent, markedly thicker and more branchy, fertile branches, sometimes emerging from anastomosing branches, resembling those in *Cladia aggregata*. In this situation it may resemble *C. rugulosa*, which is a less branched lichen with very thin central canals and very fragile podetia, or *C. signata*, a more intricately branched lichen with a somewhat felty surface, forming more or less rounded cushions and with a different chemistry. Depending on the chemistry and insolation, the colour may range from whitish grey to pale green or pale yellow.

Table 1. The 15 combinations of major chemical metabolites found in *Cladonia peltastica* in the Guianas, with numbers of specimens examined (63 in total):

I	barbatic acid (2)
II	homosekikaic acid agg. acids (1)
III	squamatic, barbatic acids (1)
IV	thamnolic acid(4)
V	thamnolic, barbatic acids (3)
VI	thamnolic, boninic acids (5)
VII	thamnolic, didymic acids (1)
VIII	thamnolic, homosekikaic agg. acids (6)
IX	usnic, barbatic acids (2)
X	usnic, homosekikaic agg. acids (1)
XI	usnic, squamatic acids (9)
XII	usnic, thamnolic acids (5)
XIII	usnic, thamnolic, barbatic acids (1)
XIV	usnic, thamnolic, ?boninic acids (7)
XV	usnic, thamnolic, homosekikaic agg. acids (15)

The report of *C. capitellata* (Hook.f. & Taylor) C. Bab. fo. *interhiascens* (Nyl.) Vain. (des Abbayes 1956: 262) as well as the reports of *Cladonia ecmocyna* G. Mey. (Meyer 1818: 297; Schomburgk 1841, misspelt as "*C. ecmozyma*") are likely to belong here.

20. **Cladonia persphacelata** Sipman & Ahti, Phytotaxa 93: 1. 2013. Type: Guyana, Upper Mazaruni Distr., Mt. Latipu, c. 8 km N of Kamarang, at c. 1000 m alt., in scrub on summit plateau, on white sand on open spot, 25 Feb 1985, H. Sipman & A. Aptroot 19149 (holotype B, isotype BRG) (TLC: thamnolic, tr. didymic acid).

– Fig. 29

Primary thallus persistent to evanescent, consisting of up to 0.5 cm long squamules which are deeply divided into c. 0.5 mm wide, elongate laciniae, attenuated and often almost stalk-like at the base, on lower side with rather smooth surface to corticoid and sometimes with ochraceous streak. Podetia up to 5 cm tall and 0.5-1.5 mm thick, of determinate growth, grey to usually more or less brown, in lower part almost black, horny and swollen, somewhat branched; branching type irregular anisotomous dichotomy, rarely trichotomy or tetrachotomy; axils closed or with usually small openings; tips often divided into 2-10 short branchlets. Podetial surface

68

Fig. 29. *Cladonia persphacelata*; type (Sipman & Aptroot 19149 (B)). Bar = 2 cm.

smooth and often shiny, denudated even at the tips, finally being rather densely squamulose, smooth inbetween, esorediate; mature squamules narrow, laciniate and imbricate, up to 4 mm long, pointing downward but with recurved tips, often glossy. Podetial wall 200-290 μm thick; cortex (0-)25-40 μm, consisting of large cells; medulla very thin, (0-)10-25 μm (including the algae); stereome distinctly delimited, very horny, thick, 200-250 μm, inner surface glossy. Conidiomata terminal on tiny apical branchlets, often grouped, 200-250 × 100-150 μm, dolioliform, constricted at base, shortly pedicellate, containing red slime. Hymenial discs not seen. Chemistry: thamnolic acid sometimes with a trace of didymic acid (TLC of 6 specimens). Colour reactions: P+yellow, K+yellow, KC-; UV-.

Distribution and ecology: A Guayana Highland endemic, known only from Venezuela and Guyana. It is widespread in the Guayana Highland of Venezuela in light, mossy forest over sandstone at c. 600-1100 m alt. In Guyana found on mossy sandstone rocks in light forest, rather shade-tolerant and avoiding open spots, from 400 to1000 m alt. Over 20 collections studied (GU: 10).

Selected specimens (paratypes): Guyana: Upper Mazaruni Distr., 2 km N of Kamarang, 500 m alt., Sipman & Aptroot 18241 (B); E-bank of Waruma R., c. 20 km S of confluence with Kako R. (campsite 4), Sipman & Aptroot 18241, 18660 (B); trail from Kamarang R. to Pwipwi Mt., c. 10 km N of Waramadan, Sipman & Aptroot 19322, 19494 (B); Potaro-Siparuni Region, Kaieteur Falls National Park, around the airstrip, Sipman 40447 (B, BRG); Region 7 (Upper Mazaruni Distr.), N of Paruima Mission, Aymatoi savanna, Sipman 39860 (B, BRG, US); Cuyuni-Mazaruni Region, Partang R., 8.6 km NE of Imbadamai, Hoffman 1722 (H).
Venezuela: Bolívar: Cerro Guaiquinima, in central part of upper plateau, along Río Carapo (near camp 3-nuevo), Sipman 27065 (B, H, VEN); in central part of upper plateau (near camp 4), Sipman 26487 (B, VEN); near NE edge of upper plateau (near camp 2), Sipman 26890 (B, VEN); near west end of upper plateau (near camp 5), Sipman 27102 (B, H, VEN); Canaima, at Río Carrao, Sipman 27256 (B, VEN).

Notes: *Cladonia persphacelata* belongs to a group of closely related species including in the Guianas *C. polystomata* and *C. subsphacelata*, and the Brazilian *C. sphacelata* Vain. *C. polystomata* grows on soil or litter and forms wide funnels on top of more or less corticated, up to c. 1 cm thick, little branched podetia with short squamules. *C. subsphacelata* has largely corticate, less than 1 mm wide podetia, and shares with *C. persphacelata* the saxicolous habit and very elongated squamules. *C. sphacelata* has short podetial squamules, the podetia remain thin, under 1 mm wide, and do not become horny.
Richly squamulose forms of *C. subdelicatula* can also resemble *C. persphacelata*. For differences see under that species.

21. **Cladonia pityrophylla** Nyl., Flora 57: 70. 1874. – *Cladonia pityrophylla* [var.] *spruceana* Vain., Acta Soc. Fauna Fl. Fenn. 10: 383. 1894, *nom. inval.* – *Cladonia pityrophylla* fo. *spruceana* Vain., Acta Soc. Fauna Fl. Fenn. 14(1): 256. 1897, *nom. inval.* Type: Brazil, Pará, Santarém, mouth of Rio Tapajos, 1849-55, Spruce, Lich. Amaz. And. 26 (holotype H-NYL 38843, isotypes BM, G, G-DC, H-NYL 38749, PC, TUR-V 19989). – Fig. 30

Primary thallus conspicuous, usually the dominant part of the lichen, consisting of imbricate, roundish, broady lobed squamules, 3-10(-15) mm long, 3.5(-10) mm wide; upper side dark green, shiny and smooth, lower side at exposed lobe ends turning violet-brown. Podetia common, laminally borne, often several per one squamule, 0.2-10 mm tall, 0.5-1.5 mm thick, of determinate growth, variegate with green, brown and grey tints, at first unbranched (often seen at this stage), later shortly branched at tips and with hymenial primordia, finally much elongated, divaricately branched and curved with developed hymenia (uncommon), usually not forming scyphi, but occasionally with narrow scyphi present at the stage of hymenial primordia. Podetial surface of sterile podetia rough, in part ecorticate, arachnoid, usually verruculose and with few or abundant flat to bullate or globose squamules; at the base with a smooth sheath of cortex extending from the primary squamules; fertile podetia much more smoothly corticate but yet verruculose, with long lateral slits and perforations. Podetial wall 160-180 μm; cortex 10-25 μm; medulla 60-80 μm; stereome 70-100 μm, sharply delimited; central canal smooth. Conidiomata scarce, stalked, on primary squamules and more commonly on either very young or older podetia, 200-250 × 100 μm, pyriform to globose, constricted at base; containg hyaline slime. Hymenial discs uncommon, their primordia (frequently with trichogynes) often abundant on corymbose branchlets at tips of larger podetia, finally solitary or forming small, dark to pale brown, 0.1-0.3 mm wide, compound discs. Chemistry: fumarprotocetraric acid and traces of protocetraric acid and confumarprotocetraric acid (TLC of 8 specimens). Colour reactions: P+ fast brick-red, K-, KC-.

Fig. 30. *Cladonia pityrophylla* (Sipman & Aptroot 18994 (B)). Bar = 2 cm.

Distribution and ecology: A Neotropical species distributed in the tropical lowlands of S America, occasionally up to 2500 m alt., from NW Venezuela to SE Brazil and Paraguay. In the Guianas found in the Pakaraima Mts. in Guyana and on an Inselberg in French Guiana. Apparently it is overlooked in the interior of Suriname, and a regular, but not abundant and inconspicuous inhabitant of open, sandy, somewhat shady places of the sandstone tablelands from 400 to 1200 m alt. Over 100 collections studied (GU: 12; FG: 1).

Selected specimens: Guyana: 8 km N of Kamarang, Mt. Latipu, Sipman & Aptroot 18994 (B); Region Potaro-Siparuni, Kaieteur Falls National Park, Ahti 52982, 52999 (H); Region 7 (Upper Mazaruni Distr.), N of Paruima Mission, Aymatoi savanna, Sipman 39919 (B, BRG). French Guiana: Mitaraka Sud, sommet inselberg, Sarthou 937 (CAY).

Note: *Cladonia pityrophylla* is an inconspicuous species with short podetia and dominant primary thallus. Characteristic are the bullate, easily detached squamules on the podetia. A stictic acid strain is known from SE Brazil.

22. **Cladonia polyscypha** Ahti & L. Xavier in Ahti *et al.*, Trop. Bryol. 7: 61. 1993. Type: Brazil, Paraíba, Mun. Alhandra, c. 30 km S of João Pessoa, 120 m alt., Ahti & Xavier Filho 45698 (holotype H, isotype JPB). – Fig. 31

Primary thallus persistent but inconspicuous, consisting of 0.5-2 mm wide, esorediate but granular squamules. Podetia 1-3(-3.5) cm tall, very slender and thin, (0.2-)0.3-0.8(-1.3) mm thick, of determinate growth, whitish to greenish grey, often becoming brownish when exposed, necrotic bases slightly melanotic, unbranched to slightly branched; tips almost regularly forming scyphi from early stage, but scyphi extremely narrow, 0.1-0.2 mm until they open up to 1(-2) mm wide at maturity. Podetial surface corticate at base or sometimes up to $\frac{1}{2}$ of the podetial length, also regularly inside and outside fertile scyphi and often on basal parts of the branchlets; otherwise sorediate, soredia either farinose or immixed with granules or isidioid microsquamules (0.1-0.2 mm long) or tiny squamules. Podetial wall 160-230 μm; cortex 10-25 μm; medulla 50-75 μm; stereome 100-125 μm; surface of central canal densely papillulate. Conidiomata common, terminal, globose to pyriform, clearly constricted at base, 0.8-1.5 × 1-2.1 mm. Hymenial discs brown, globose, single or in glomerule-like groups. Chemistry: fumarprotocetraric acid and traces of protocetraric and confumarprotocetraric acids (TLC of 22 specimens). Colour reactions: P+ red, K-, KC-.

72

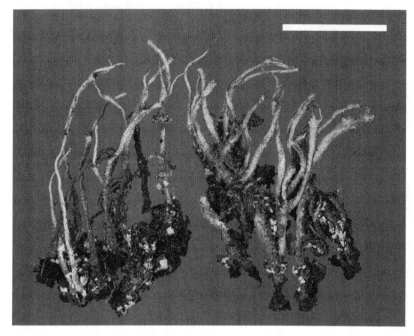

Fig. 31. *Cladonia polyscypha* (Sipman 40309 (B)). Bar = 2 cm.

Table 2. Differences between *Cladonia subradiata* and *C. polyscypha*.

Cladonia subradiata	*Cladonia polyscypha*
podetia ecorticate, except sometimes at base	podetia more or less corticate at base and scyphi
colour whitish, not melanotic	colour darker, even brownish, slightly melanotic
(almost) unbranched	more branched when old
rarely forming scyphi	often forming scyphi, even on very thin podetia
rarely fertile	regularly fertile
pycnidia scarce, pale brown, few	pycnidia black, numerous
mainly on decaying wood	mainly on sand
quaesitic acid (Cph-1) constant	quaesitic acid (Cph-1) not constant
not robust	generally more robust

Distribution and ecology: Endemic to S America, where it seems to occur mainly in the Guianas with adjacent Venezuela and the Brazilian coast. Widespread in the Guianas, growing often in abundance on sandy soil in open spots in scrub savanna or secundary vegetation, from 50 to 400(-1150) m. As substrate it uses more or less horizontal sand, abandoned termite nests, more rarely rotten logs. Over 150 collections studied (GU: 96; SU: 12; FG: 14).

Selected specimens: Guyana: Demerara R., Jenman 7377 (BM); Pakaraima Mts., Mt. Aymatoi, Maas et al. 5842 (H). Suriname: Dam, Florschütz & Florschütz 176 (H). French Guiana: St Elie, Sastre 513 (H); Montsinéry, Aptroot 15119 (H).

Notes: This is the commonest *Cladonia* species in the Guianas and it is apparently favoured by human activities. Because of its mostly subulate, sorediate podetia, it could be confused with certain forms of *C. subdelicatula*, which have a different chemistry.

C. polyscypha is very similar to *C. subradiata*, a widespread and common species on mossy banks and rotten logs throughout the tropics. They share the subulate, more or less sorediate podetia which may produce microsquamules and may form small scyphi, the brown hymenial discs and the chemistry. The differences are mainly gradual or only present in certain stages (Table 2) and it could be concluded that both species are virtually indistinguishable, except when in well developed stages. This is a common situation in the genus *Cladonia*, but often chemical markers help to overcome the lack of morphological characters. In the present case, there are unfortunately no such chemical markers to recognize morphologically poor material. All material sufficiently well developed from the Guianas showed the characters of *C. polyscypha*, and therefore it is possible that all material from the complex belong here.

Reports of *Cladonia ramulosa* (With.) J.R. Laundon and its synonyms *C. pityrea* (Flörke) Fr. var. *ramosa* Flot. ex Hampe, *nom. nud.* (Schomburgk 1849: 1041) and *C. pityrea* "fo. *cladomorpha* Flörke" (Hue 1898: 278) probably belong here.

23. **Cladonia polystomata** Ahti & Sipman, Fl. Neotrop. Monogr. 78: 294. 2000. Type: Brazil, São Paulo, Mun. Itanhaém, c. 7 km WSW of Itanhaém, km 2 along turnoff to NW from km 115.5 on Hwy. SP-56, 10 m alt., Ahti, Marcelli & Vitikainen 55633 (holotype SP 158578, isotypes CTES, H, IBUG, NY, US). – Fig. 32

Primary thallus persistent to evanescent, consisting of 1.5-3 mm long and 0.25-0.5 mm wide squamules. Podetia 1-3 cm tall, 0.5-1 mm thick, of determinate growth, whitish-grey with brownish base, somewhat

74

Fig. 32. *Cladonia polystomata* (Sipman & Aptroot 19138 (B)). Bar = 2 cm.

branched by dichotomy or polytomy; axils open and often widened to about three times the podetium width, giving the podetia a slightly scyphoid habit; tips subulate, usually up to 1 mm long and on the margin of widened, open axils. Podetial surface almost totally ecorticate, beset near the top with corticate warts, lower down squamulose, esorediate; squamules patent, with upturned tips. Podetial wall 120-150(-225) μm; cortex (0-)10-15 μm; medulla 30-50 μm; stereome 120-140 μm, hard; surface of central canal papillate. Conidiomata terminal, 130-160 μm, cylindric, not constricted at the base, containing hyaline slime. Hymenial discs frequent, 0.1-0.5 mm wide, forming corymbose groups at podetial tips of slightly thicker podetia with smooth surface, blackish-brown when mature. Chemistry: thamnolic acid, traces of decarboxythamnolic acid and often additional, unknown, minor compounds (TLC of 2 specimens). Colour reactions: P+ yellow, K+ yellow, KC-.

Distribution and ecology: Apparently endemic to S America. The range of this species seems to be bicentric: most records are from SE Brazil but a large outlier exists in Guyana and Venezuela. In the Guianas so far known from 4 specimens from the Pakaraima Mts. in Guyana, where it occurs on sand in open places in scrub savanna, from 450 to 1000 m alt. Over 50 collections studied (GU: 5).

Specimens examined: Guyana: Upper Mazaruni Distr., 8 km N of Kamarang, Mt. Latipu, Sipman & Aptroot 19137 (B, BRG), 19138 (B, BRG, H); vic. Kamarang, Boom & Gopaul 7623 (BRG, NY); Trail from Kamarang R. to Pwipwi Mt., c. 5 km N of Waramadan, Sipman & Aptroot 19251c (B, BRG).

Notes: *Cladonia polystomata* resembles the widespread *C. squamosa*, a species from cold environments, but its podetia are very slender and its squamules are very small. In the Guianas it is most easily confused with *C. persphacelata* and *C. subdelicatula*. For differences see under these species.

24. Cladonia prancei Ahti, Fl. Neotrop. Monogr. 78: 223. 2000. Type: Brazil, Amazonas, Basin of Rio Negro, along Camanaus-Uaupés rd., near Camanaus, Prance *et al.* 15944 (holotype INPA, photo H, NY, isotypes H, NY).   – Fig. 33

Primary thallus squamulose to crustose, consisting of small, 1-2 mm long and 0.2-1 mm wide, laciniate, ascending squamules, with convex upper side, grey to orange-yellow in necrotic parts (and hypothallus), not white-maculate, at the base fairly compact but ecorticate and fibrose, esorediate or almost totally dissolved into isidioid soredia, connected with yellow-orange hypothallus. Podetia up to 3 cm tall, 0.5-4 mm thick, of determinate growth, sordid white to pale yellowish grey, digitately branched; typically at first (up to 1.5 cm tall) unbranched, clavate, somewhat swollen (widest above the base), at later stage 2-4 times or more successively branched, sometimes richly branched, branching irregularly dichotomous, axils closed; tips subulate or blunt to scyphi-shaped; scyphi narrow, 1-2 mm wide, margins incurved and often strongly proliferating. Podetial surface abundantly farinose-sorediate throughout, soredia immixed with tiny squamules in basal parts, sometimes a very short corticate sheath present at base. Podetial wall 160-300(-350) µm thick; medulla fully sorediate,

Fig. 33. *Cladonia prancei* (Duivenvoorden et al. 199c (B)). Bar = 2 cm.

(0-)100-175 μm thick, including the discontinuous algal colonies; stereome 160-200 μm, strong, compact, pellucid, sharply delimited from medulla; inner surface glossy, uneven. Hymenial discs uncommon, red, forming up to 2 mm wide compound discs. Conidiomata terminal, subterminal or lateral (near tips) on podetia, spherical to shortly ovoid, 250-300 μm diam., often with short stalk, not or little constricted at base, black to brown, containing purple slime. Chemistry: thamnolic acid, sometimes with didymic acid (TLC of 3 specimens). Colour reactions: P+ yellow, K+ yellow, KC-.

Distribution and ecology: Essentially this is an Amazonian lowland species, extending to Peru and Ecuador, and to lower elevations of the C Brazilian mountains. Several specimens are available from throughout Guyana and Suriname, from scrub savanna on sandstone tableland or on white sand, mostly growing on logs, stumps or tree boles along rivers, but also on soil, e.g. on white sand, from 0 to 800 m alt. Over 25 collections studied (GU: 8; SU: 3).

Selected specimens: Guyana: Kamarang, Boom & Gopaul 7244 (BRG, NY); Region Potaro-Siparuni, Kaieteur Falls National Park, Ahti 53052, 53187 (H); id., Paramakatoi, Ahti 53357a (H). Suriname: Brinckheuvel, Teunissen & Wildschut LBB 11933 (H, L); Marieposa, Kegel 1448 (GOET).

Notes: *Cladonia prancei* is characterized by its conspicuous, pale whitish to yellowish grey, abundantly sorediate, subulate to narrowly scyphoid podetia. It is close to *C. corallifera* and *C. mollis* by the form of its squamules and the yellow-orange hypothallus.
Outside the Guianas also usnic, squamatic and barbatic acid are found in the species

25. Cladonia pulviniformis Ahti, Lichenologist 22: 263. 1990. Type: Venezuela, Bolívar, Dist. Piar, Macizo del Chimantá, Acopán-tepui, 2250 m alt., Ahti et al. 45045 (holotype VEN, isotypes B, COL, DUKE, H, MYF, NY, US). – Fig. 34

Primary thallus not seen. Podetia 5-12 cm tall, of indeterminate growth, ash-grey to blackish-brown, at least apical parts easily browned, forming very densely branched cylindrical cushions 2-7 cm wide, which can be more or less coherent and form large mats several decimeters across; branching type subisotomous dichotomy, with some trichotomy; no main stem but occasionally some stronger branches (0.4-0.6 mm wide) distinguishable; axils usually closed, sometimes perforated (especially when branching trichotomous). Podetial surface verrucose, fibrose,

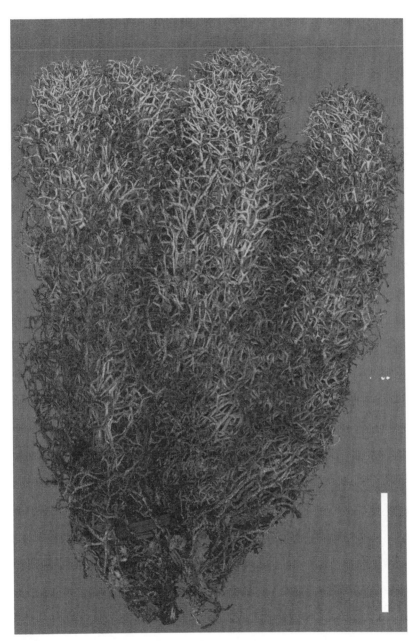

Fig. 34. *Cladonia pulviniformis* (Sipman 40310 (B)). Bar = 2 cm.

ecorticate; squamules never observed. Podetial wall 160-200 μm, soft and weak, without cortex; stereome replaced by a layer of incompletely to almost completely conglutinated hyphae; surface of central cavity accordingly variable, felty to papillulate, striate. Conidiomata 200-250 × 100-150 μm, containing red slime. Hymenial discs not seen. Chemistry: barbatic acid, usually with thamnolic acid (TLC of 22 specimens). Colour reactions: P- or + yellow, K- or + yellow, KC-.

Distribution and ecology: Known only from the Guiana Shield in Colombia and Venezuela, from lowland savannas at 200 m alt. near Araracuara up to paramoid vegetation above 2000 m alt. in the Guayana Highlands. In the Guianas, mainly known from the sandstone tablelands in the Pakaraima Mts. of Guyana, more occasional in lowland white sand savannas of Guyana and Suriname, from 50 to 1200 m alt. On the tablelands it can be an important component of the lichen vegetation on open places in scrub savanna on rock plateaux. Over 100 collections studied (GU: 48; SU: 5).

Selected specimens: Exsicc.: Lichenotheca Latinoamericana 60. Guyana: Region Potaro-Siparuni, Kaieteur Falls National Park, Kvist 232 (BRG, H, L); Upper Demerara-Berbice Region, 5 km E of Rockstone, Pipoly 9662 (H); Upper Mazaruni Distr., Mt. Latipu, Sipman & Aptroot 19151 (B); 10 km N of Waramadan, Sipman & Aptroot 19329. Suriname: Jodensavanne, Kegel 996 (GOET); Zanderij, near LBB school, Lindeman & Mennega *et al.* 109 (L).

Notes: Usually *C. pulviniformis* is easily identified due to its extremely dense, frequently dark brown, columnar podetia. These are very soft because they have no stereome and only rudimentary skeletal tissue, which is composed of poorly conglutinated hyphae and thus easily split longitudinally and felty at the ruptures.
The species can be confused with *C. huberi*, a rare species from bog in the upper parts of the Guayana Highlands, and it is also similar to *C. variegata*. For differences, see under those species.
Outside the Guianas 3 further chemotypes are reported: (II) thamnolic acid only, (III) barbatic acid plus four unidentified minor compounds, and (IV) squamatic acid.

Cladonia rangiferina (L.) F.H. Wigg. subsp. **abbayesii** (Ahti) Ahti & DePriest, Mycotaxon 78: 501. 2001. – *Cladonia rangiferina* var. *abbayesii* Ahti, Ann. Bot. Soc. Zool. Bot. Fenn. Vanamo 32(1): 94. 1961. – *Cladina rangiferina* (L.) Nyl. subsp. *abbayesii* (Ahti) W.L. Culb. in Vězda, Lich. Sel. Exs. Fasc. 76: 3 [no. 1883]. 1983. Type: Colombia, Cundinamarca, Bogotá, 3100 m alt., Lindig 2513 (holotype UPS, isotypes BM, G, H-NYL p.m. 1184, M, REN, S).   – Fig. 35

Fig. 35. *Cladonia rangiferina* ssp. *abbayesii* (Harris 15429 (B)). Bar = 2 cm.

Primary thallus absent. Podetia up to 12 cm tall, of indeterminate growth, pale grey to bluish-grey, often strongly browned over c. 0.2-0.5 mm at the apical tips and with brownish branchlets near the tips, dead basal parts turning dark brown but hardly black, rather loosely branched, with c. 3-5 mm long internodes when mature, forming flat, dense colonies with narrow (up to 1.5 cm wide) or no podetial heads; predominant branching type anisotomous dichotomy but trichotomy also common; main axes distinct, often robust (1-2 mm thick at base); axils open or closed; ultimate apical branchlets relatively thick and blunt, falcate to erect, forming narrow heads. Podetial surface smooth near tips, more or less verrucose towards the base, clearly arachnoid. Podetial wall 100-400 μm, ectal layer up to 60 μm; stereome 100-240 μm. Conidiomata containing hyaline slime. Chemistry: atranorin and fumarprotocetraric acid, with traces of protocetraric acid, confumarprotocetraric acid, sometimes ursolic acid, and unidentified fatty acids (Ahti 2000). Colour reactions: P+ orange red, K+ yellow.

Distribution and ecology: A Neotropical taxon largely restricted to the humid Andean paramos (2450-4050 m), with a few records from the Guayana Highlands in adjacent Venezuela. The species is more widespread in the temperate and cold zones of the Northern Hemisphere. Not yet known from the Guianas. Over 100 collections studied.

Selected specimen: Venezuela: Bolivar, Macizo del Chimantá, Ahti et al. 44986 (B, H).

Notes: *Cladonia rangiferina* subsp. *abbayesii* is most likely to be confused with *Cladonia sprucei*. For differences, see under that species. A record of the species from Guyana (Ahti 2000) is referred here as *C. sprucei,* because the specimen contains psoromic acid as minor compound, a substance unknown from *C. rangiferina* but locally common in *C. sprucei.*

Cladonia rappii A. Evans, Trans. Connecticut Acad. Arts 38: 297. 1952. Type: U.S.A., Florida, Seminole Co., Sanford, 1924, Rapp in Sandstede: Cladon. Exs. 1938 p.p. (holotype US, isotypes NY, TUR-V 18217, UPS). – Fig. 36

Primary thallus mostly persistent, consisting of 2-4 mm long and 2 mm wide, elongate, sparsely lobed, often incurved squamules. Podetia slender to rather robust, 3-6 cm tall, 0.5-1.5 mm thick, usually of determinate growth, pale grey to dark brownish, usually strongly melanotic at the

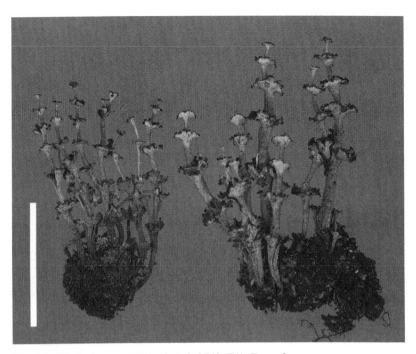

Fig. 36. *Cladonia rappii* (Cleef et al. 3540 (B)). Bar = 2 cm.

base, sparsely branched, composed of centrally proliferating scyphi in 3-12 tiers; scyphi 3-14 mm long, the basal one sometimes longer, up to 25 mm, abruptly flaring, shallow, 2-4 mm wide, not perforated or with some slits in old podetia; scyphal margins upturned, soon denticulate to slightly divided, much dissected when abundantly fertile. Podetial surface areolate-corticate, sometimes squamulose or scaling off to expose the stereome, minutely pruinose-arachnoid, especially in young parts. Podetial wall 300-340 μm thick; cortex 24-28 μm; medulla 140-200 μm; stereome 100-140 μm, distinctly delimited, soon darkening and finally blackening, rather soft; surface of the central canal papillose. Conidiomata 0.2-0.25 × 0.2-0.25 mm, sessile to short-stalked, on scyphal margins, occasionally on the bottom of the scyphi, globose to shortly pyriform, constricted or not at base, containing hyaline slime. Hymenial discs 1-1.5(-3) mm wide, on short, flat stalks on the margins of the scyphi, long remaining flat on top but finally convex, dark to pale brown. Chemistry: fumarprotocetraric acid and traces of protocetraric, quaesitic (subconstant) and confumarprotocetraric acids, and other unknown minor components (Ahti 2000). Colour reactions: P+ orange red, K-, KC-.

Distribution and ecology: A widespread species occurring in N America, Asia, Africa and Melanesia. In the Neotropics it is a high-altitude species, common along the Andes and in SE Brasil, and sporadically found around 2000 m alt. in the Venezuelan part of the Guayana Highlands. Specimens were available from Kavanayén, Ueitepui, Chimantá and Auyántepui in Bolívar. Not yet known from the Guianas. Over 200 collections studied.

Notes: *Cladonia rappii* is rather variable so that further studies may reveal that it contains more than one taxon. One specimen from Venezuelan Guayana (Huber 10046) is particularly deviating by its numerous short (3-4 mm) internodes.
From elsewhere known are chemical strains with psoromic acid instead of fumarprotocetaric acid (with colour reaction P+ yellow), and with additional atranorin (colour reaction K+ pale yellow).

26. **Cladonia recta** Ahti & Sipman, Phytotaxa 93: 1. 2013. Type: Guyana, Upper Takutu Distr., c. 35 km S of Aishalton, c. 5 km N of Kuyuwini Landing, along track to Karaudanawa, c. 2° 08' N, 95° 15' W., c. 250 m alt., on sandy soil among scattered shrubs and trees along and on small savanna, 1 Nov. 1992, H. Sipman 57134 (holotype B 60 0164014, isotype BRG). – Fig. 37

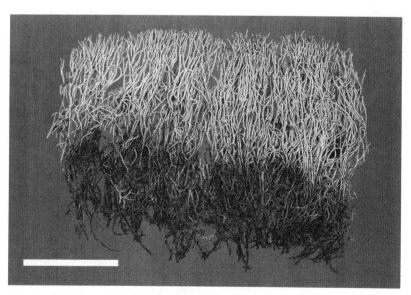

Fig. 37. *Cladonia recta*; type (Sipman 57134 (B)). Bar = 2 cm.

Primary thallus evanescent, consisting of very small, to. 0.2 mm long, slightly crenulate squamules. Podetia 2-5 cm tall, 0.2-0.5(-1) mm thick, of indeterminate growth, whitish grey, basal parts darker but not melanotic, extreme tips black, forming dense mats with individual podetia erect and very straight, branched by irregular anisotomous dichotomy, rarely trichotomy, main stems distinct but somewhat anastomosing, equally thick; axils normally closed but occasionally perforated; tips erect or slightly bent, acuminate. Podetial surface continuously corticate, cortex smooth or somewhat rugulose, especially towards the base, occasionally slightly squamulose, dull to slightly shiny, somewhat maculate, with some brownish, ecorticate patches near the base; soredia lacking. Podetial wall anatomy not studied; stereome distinct, central canal very narrow. Conidiomata on tips of podetia, c. 200 × 100 μm, cylindrical, not constricted at the base, black, containing purple slime. Hymenial discs at tips of slightly swollen podetia, very small (c. 0.1 mm diam.), pale brown; spores not observed. Chemistry: squamatic acid. Colour reactions: P-, K-, KC-.

Distribution and ecology: As far as known a Guianan endemic, collected in Guyana on sandy soil and rotten wood in savanna, at c. 250 m alt. (GU: 3).

Additional specimens (paratypes): Guyana: Upper Takutu Distr., c. 35 km S of Aishalton, c. 5 km N of Kuyuwini Landing, along track to Karaudanawa, c. 2° 08' N, 95° 15' W, c. 250 m alt., Sipman 57127, 57132 (B, BRG).

Notes: *Cladonia recta* resembles most *C. peltastica*, but is much more slender and forms conspicuous, straight, erect colonies. However, it is possible that it represents an usnic acid-free chemotype of that species.

27. **Cladonia rotundata** Ahti, Ann. Bot. Soc. Zool. Bot. Fenn. Vanamo 32(1): 29. 1961. – *Cladina rotundata* (Ahti) Ahti, Beih. Nova Hedwigia 79: 40. 1984. Type: Peru, San Martín, Roque, La Campana, Melin 7 (holotype UPS, isotype S). – Fig. 38

Primary thallus absent. Podetia 8-10 cm tall, of indeterminate growth, pale ash-grey or pale yellowish-green, the exposed parts usually distinctly browned, with mature internodes usually 2 mm long, 0.4-0.8 mm thick, in part slightly flattened, forming dense, semiglobose heads, which usually fuse to form large colonies; branching type fairly regular isotomous dichotomy; ultimate branchlets fairly coarse and strongly divergent; main

84

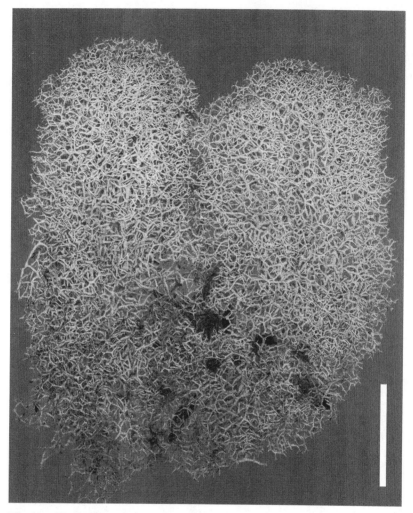

Fig. 38. *Cladonia rotundata* (Sipman 26721 (B)). Bar = 2 cm.

stems absent or very poorly differentiated; base not melanotic. Podetial surface with very thin arachnoid cover near the tips, soon glabrous, verruculose by the protuberant algal cell clusters. Podetial wall very thin, 70-80 μm. Conidiomata containing hyaline (or sometimes red?) slime. Hymenial discs not seen. Chemistry: atranorin and fumarprotocetraric acid, with traces of protocetraric and confumarprotocetraric acids, and sometimes also of usnic acid (Ahti 2000). Colour reactions: P+ fast red, K+ yellow, KC- or KC+ yellow.

Distribution and ecology: A neotropical species known only from the Amazon, from Colombia and Brazil to Peru, growing on sand in savannas (often in considerable shade under shrubs) and in cloud forests up to 2400 m alt. In the Guianas, it is known from the Pakaraima Mts. of Guyana and from a single locality in Suriname. Over 20 collections studied (GU: 6; SU: 2).

Selected specimens: Guyana: Region 7 (Upper Mazaruni Distr.), N of Paruima Mission, Aymatoi savanna, Sipman 39878 (B, BRG); Mt. Aymatoi, DePriest 10461 (BRG, H, US). Suriname: Zanderij, Florschütz & Florschütz 759, 765 (L).

Notes: *Cladonia rotundata* is a relatively little known species, most similar to *C. dendroides*. It can be distinguished by its strong isotomy, the absence of blackening at the base, and the hyaline slime in the conidiomata. Earlier (e.g., Ahti 1961, and in herbarium annotations) it was erroneously included in *C. dendroides*.
In its shape it is very similar to *C. confusa*, and particularly shade forms of the latter can be easily confused. They have a clearly different chemistry, however.

28. **Cladonia rugulosa** Ahti, Ann. Bot. Fenn. 23: 206. 1986. Type: Venezuela, Mérida, Pico de Horma area, Morro Negro, SE of Mesa Quintero, 2400 m alt., López-Figueiras & Rodríguez 23086 (holotype MERF, isotypes H, US, VEN).   – Fig. 39

Primary thallus persistent, grey, consisting of 2 mm long and 1 mm wide squamules, with narrow lobes, lower side arachnoid but fairly compact. Podetia very fragile, 1.5-6 cm tall, 0.2-0.6 mm thick, of indeterminate growth, lower parts darkgrey (in Venezuelan populations whitish grey, rarely greenish-yellow), upper parts pale grey (in Venezuelan populations variegated, with grey verruculae and brown interspaces), unbranched or 1-3 times dichotomously branched, occasionally anastomosing; axils usually closed, rarely perforated; forming dense, erect, fasciculate colonies; mature hymenia-bearing podetia much more robust, 0.5-0.7(-2.0) mm thick and corymbosely branched at apex, with tips without scyphi. Podetial surface almost smoothly corticate to slightly rugulose, in Venezuelan populations upper parts clearly rugulose, esorediate; podetial squamules small, scarce, confined to the very base. Podetial wall 135-165 μm thick; cortex 7-10 μm; medulla 35-65 μm; stereome 50-100 μm, well-limited, horny and fragile; central canal very narrow, papillose or slightly fibrose on surface; the thinnest distal parts of podetia completely solid. Conidiomata terminal on podetia, always solitary, cylindrical, 0.4-0.5 × 0.1 mm, often stalked but hardly

Fig. 39. *Cladonia rugulosa* (Sipman 40317 (B)). Bar = 2 cm.

constricted at base; containing hyaline slime. Hymenial discs at tips of short stalks at the apices of swollen, branchy podetia, dark brown. Chemistry: thamnolic and (in hymenial region) barbatic acids, also traces of decarboxythamnolic acid, sometimes boninic acid and unknown compounds, and occasionally of usnic acid (TLC of 9 specimens). Colour reactions: P+ yellow-orange, K+ yellow-orange, KC-.

Distribution and ecology: An endemic of northern S America centered in the Guayana Highland, locally common in the Venezuelan part at high elevations (above 2500 m), also known from the Venezuelan Andes. In the Guianas it is known from the sandstone tableland of the Pakaraima Mts. in Guyana, where it is regularly found in scrub savanna on rock plateaux with scarce vegetation, on spots which remain long wet by water trickling from the surrounding vegetation, from 400 to1000 m alt. Over 50 collections studied (GU: 25; SU: 1).

Selected specimens: Guyana: Region Potaro-Siparuni, Kaieteur Falls National Park, DePriest 9291 (H); 10 km N of Waramadan, Sipman & Aptroot 19335 (B, H); Mt. Latipu, 8 km N of Kamarang, Sipman & Aptroot 19144 (B); Region 7 (Upper Mazaruni Distr.), N of Paruima Mission, Aymatoi savanna, Sipman 39815 (B, BRG); Region Cuyuni-Mazaruni, Pakaraima Mts., along Partang R., 5 km SE from Imbaimadai settlement, Henkel *et al.* 6 (B, US).

Notes: *Cladonia rugulosa* is an inconspicuous lichen usually growing in scattered, small clumps, characterized by little branched, whitish, very fragile, erect podetia with very narrow central canal. By its colour, it resembles *C. hians*, a species with much wider podetia with open, dilatated axils. It is probably closest to *C. peltastica*, which is more densely branched and not fasciculate. Ahti (2000) related it to *C. mutabilis* Vain. and *C. polytypa* Vain., but these species are much more distinct due to their wider podetia with swollen, open axils.
The population of Guyana deviates considerably from the Venezuelan material by its 1) larger size, the podetia measuring 3-6 cm tall and 4-6 mm diam. instead of 1-2 cm tall and 3-4 mm diam.; 2) podetia with smooth and faintly variegated instead of grey and brown-variegated upper parts. The material from Guyana has often a yellow medulla (pigment K-), which is visible on broken spots as yellowish vessels, while in Venezuelan material there is never such a yellow medulla, and no vessels are exerted upon breaking of the podetia. It remains to be seen whether these are properties restricted to the one locality where most specimens are from (Kaieteur Falls), or whether these characters mark a more widely distributed population.

29. **Cladonia rupununii** Ahti & Sipman, Phytotaxa 93: 1. 2013. Type: Guyana, Upper Takutu Distr., southern Rupununi savanna, Kusad Mt., SE side, 2° 47' N, 59° 51' W, 450 m alt., forest along stream in upper part of valley, on *Curatella* trunk, 29 Sept. 1992, H. Sipman 57749 (holotype B, isotype BRG). – Fig. 40

Primary thallus persistent, consisting of very small, 0.5-2 mm long, dissected squamules, veined below; often slightly sorediate along margins. Podetia up to 2 cm tall, 0.3-1 mm thick, of determinate growth, very slender, whitish grey, dichotomously or digitately branched, especially in upper parts; branchlets divergent; axils closed; tips bluntish; acysphose. Podetial surface abundantly farinose-sorediate throughout. Podetial wall anatomy not studied. Hymenial discs red, not seen in maturity. Conidiomata born on basal squamules or at tips of podetia, ampullaceous, black, containing purple slime. Chemistry: didymic, barbatic, trace of demethylbarbatic acids. Colour reactions: P-, K-, KC-.

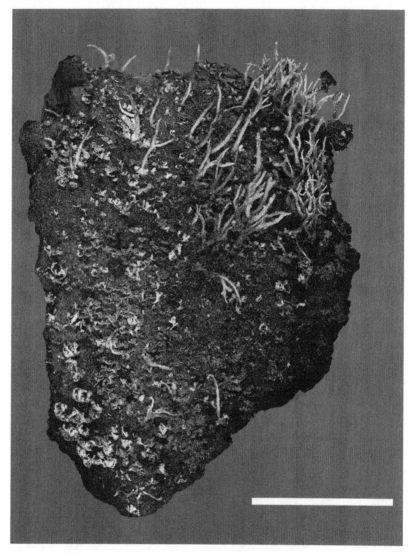

Fig. 40. *Cladonia rupununii*; type (Sipman 57749 (B)). Bar = 2 cm.

Distribution and ecology: A Guyana endemic, so far as known, collected in a forest island on a hilltop in the Rupununi savanna in SW Guyana, on rotten log in clearing along a stream, at 450 m alt. One collection studied (GU: 1).

Selected specimens: type only.

N o t e s : Very similar and apparently closely related to *Cladonia macilenta* Hoffm., but more branched. Also similar to *Cladonia prancei*, which is much more robust, forms narrow scyphi, and always contains thamnolic acid. Chemically, it is similar to *C. didyma*, but that species has smooth, pellucid podetia producing granules or squamules and lacks soredia.

30. **Cladonia secundana** Nyl., Flora 57: 71. 1874. – *Cladonia miniata* [fo.] *secundana* (Nyl.) Vain., Acta Soc. Fauna Fl. Fenn. 4: 67. 1887. – *Cladonia miniata* var. *secundana* (Nyl.) Zahlbr., Catal. lich. univ. 4: 561. 1927. Type: Venezuela, Amazonas, Depto. Río Negro, San Carlos de Río Negro, Spruce, Lich. Amaz. And. 35 (lectotype H-NYL 37826, designated by Stenroos 1989a, isolectotypes BM, G).
– Fig. 41

Cladonia erythromelana Müll. Arg., Flora 65: 298. 1882. – *Cladonia miniata* var. *erythromelana* (Müll. Arg.) Zahlbr., Catal. Lich. Univ. 4: 561. 1927, as '*erythromelaena*'. Type: Brazil, State Rio de Janeiro, Glaziou 12327 (lectotype G, designated by Ahti & Stenroos 1986, isolectotypes BM, FH-Taylor, G, H, LE, M, TUR-V 14194, UPS).

Primary thallus persistent and often dominant, consisting of (1-)2-7(-10) mm long, 0.25-0.4 mm thick, deeply lobed squamules with 0.8-1.5 mm wide, often linear lobes, entire margins and blunt tips, imbricate but not showing conspicuous hygroscopic movements when wetted; lower surface whitish, yellowish or bluish grey, often longitudinally veined, scarcely melanotic;

Fig. 41. *Cladonia secundana* (Sipman 40319 (B)). Bar = 2 cm.

cortex 100-200 µm, hard-cartilaginous, translucent; emorient base slightly orange, but rhizoids (well-developed) mostly white. Podetia borne laminally or marginally (becoming phyllopodial) on the primary squamules, up to 1(-2) cm tall, c. 1(-3) mm thick, of determinate growth, olive green to brown, subcylindrical, 1-3 times branched from near the base or with terminal clusters of fine branchlets. Podetial surface continuously corticate, smooth to rugulose and cracked; sometimes with large squamules. Podetial wall 300-400 µm thick; cortex 50-120 µm; medulla 50-400 µm; stereome 30-150 µm, continuous to subcontinuous, somewhat stranded; central canal grooved and rather coarsely papillose. Conidiomata 0.2-0.3 mm long, 0.15-0.2 mm wide, black to red; borne laminally or marginally on the primary squamules or terminally on the podetia, not constricted at base but often with short, broad stalk; containing red slime. Hymenial discs commonly produced on the podetium tips, red, 0.5-2(-3) mm wide. Chemistry: didymic acid, often with traces of unidentified substances, once with thamnolic acid (TLC of 12 specimens). Colour reactions: P-, K-, rarely + yellow to red, KC-.

Distribution and ecology: Neotropical endemic species known from the Guayana Highlands throughout Amazonia to SE Brazil, distributed in the lowlands and lower to middle elevations (to c. 2500 m) of mountains. In the Guianas it is rather widespread and relatively common in Guyana, once found in Suriname, and not yet known from French Guiana. It occurs mostly on sandy soil on open places in the white sand savannas and sandstone tablelands and may also be found on humous soils, bare peat in bogs, and rotting logs, from sea level to1000 m alt. Over 100 collections studied (GU: 36; SU: 3).

Selected specimens: Guyana: Potaro-Siparuni Region, Kaieteur Falls National Park, Hahn et al. 4134 (US), 4682 (H); Pakaraima Mts., Mt. Latipu, Maas & Westra 4198 (H, L); Region Demerara-Mahaica, Linden Highway, DePriest 9238 (H); Cuyuni-Mazaruni Region, Haiamatipu, McDowell et al. 4673 (US), trail from abandoned balata bleeders camp at base of Mt. Makarapan to Rupununi R., Maas et al. 7576 (L); Region 7 (Upper Mazaruni Distr.), N of Paruima Mission, Aymatoi savanna, Sipman 39839 (B, BRG). Suriname: Nickerie district, area of Kabalebo Dam project, at Km 96, Dakama-savanna, Zielman 1473 (BBS, L).

Notes: A species of the *C. miniata*-group unlikely to be confused with other species in the Guianas except *C. guianensis*. The latter differs by having shorter and wider, often inconspicuous, primary thallus lobes, much shorter than the podetia, and by the squamulose podetia, usually over 1 cm tall and scarcely branched. Forms of *C. secundana* with well developed podetia and poorly developed thallus differ by the very branchy, not squamulose podetia.

Considering its whole range, the secondary chemistry is extremely variable. According to Stenroos (1989a) the major medullary compounds are didymic acid (with condidymic and subdidymic acids; in 52% of studied specimens), barbatic acid (24%, in part as minor compound) or sekikaic and homosekikaic acids (17%). The accessory and trace substances include 4-O-demethylbarbatic, 3-a-hydroxybarbatic, squamatic, consquamatic, thamnolic, usnic, 4'-O-methylnorsekikaic, 4'-O-methylnorhomosekikaic, fumarprotocetraric, protocetraric and confumarprotocetraric acids, atranorin, as well as the unknown substances M1, M2, M3, M6, M7, M9, M11, M13, M14, M19, M29 and M30. The presence of 30 recognized secondary compounds is probably the highest number recorded in any species of *Cladonia*. The different chemotypes have apparently no distinct geographical ranges, but occur more or less at random.

Reports of *Cladonia miniata* G. Mey. var. *anaemica* (Nyl.) Vain. (Des Abbayes 1956: 259) probably belong here.

31. **Cladonia signata** (Eschw.) Vain., Medd. Soc. Fauna Fl. Fenn. 14: 32. 1886. – *Cladonia rangiferina* [var.] *signata* Eschw. in Mart., Fl. Bras. 1(1): 275. 1833. Type: Brazil, Amazonas, "In sylviis fluvii Amazonum" (not found, probably destroyed in B). Neotype: Brazil, Amazonas, Rio Cuieiras, 50 km upstream, 1974, Ongley & Ramos P21767 (holoneotype INPA, designated by Ahti 1993, isoneotypes DUKE, H, M, NY, US).   – Fig. 42

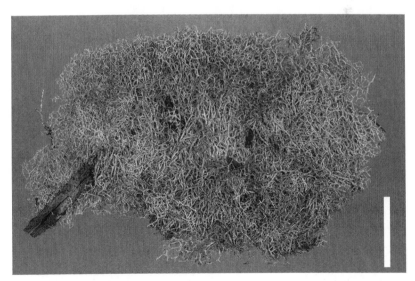

Fig. 42. *Cladonia signata* (Sipman 40324 (B)). Bar = 2 cm.

Primary thallus very rarely seen (not in Guianas material), consisting of 0.1-0.2 mm wide, soft, scarcely corticate squamules. Podetia 2.5-12 (-20) cm tall, 0.3-0.6(-0.8) mm thick, of indeterminate growth, pale greenish-grey, but usually strongly browned in upper parts (commonly variegated), not melanotic at base, densely branched by anisotomous dichotomy or to a lesser degree trichotomy, forming wide, continuous, flattish mats, in part with irregular rounded heads, almost without distinguishable main stems, with strongly anastomosing branches; axils usually closed; tips erect, divaricate, in part slightly flattened, subulate. Podetial surface ecorticate, subarachnoid to entirely glabrous; esorediate; very rarely with squamules; squamules very small, c. 0.1 mm wide, rounded to crenulate, only seen at very base. Podetial wall very thin, 60-160 μm; cortex absent, medulla (0-)10-60 μm; stereome 60-100 μm; central canal smooth. Conidiomata 90-110 μm thick, cylindrical to dolioliform, not or slightly constricted at base; containing hyaline slime. Hymenial discs 0.3-1 mm wide, arranged in corymbose manner at podetial tips, peltate, dark brown. Chemistry: fumarprotocetraric acid (major; 1.3-3.0% of dry weight) with homosekikaic acid (minor, inconstant), probably hyperhomosekikaic acid (major, inconstant), confumarprotocetraric acid (minor), and minor unknowns, or with atranorin (TLC of 30 specimens). Colour reactions: P+ fast red, K- (or brownish)/+ pale yellow, KC-.

Distribution and ecology: Endemic to tropical S America, widespread and often common in the lowlands, in Amazonia, Guayana Highlands, SE Brazil and along the Andes from Venezuela and Colombia to Bolivia. In the Guianas, widespread in Guyana and Suriname, from (0-)300 to 1000 m alt., growing on the ground in light forest on humid, acid and oligotrophic places. It is common in the white sand areas, but inhabitting rather shady or even moist (swampy) habitats along the margins of the sand patches. Over 300 collections studied (GU: 41; SU: 9).

Selected specimens: Exsicc.: Lichenotheca Latinoamericana 11, 12, 62. Guyana: Region U. Demerara-Berbice,15 km E of Rockstone, Pipoly 9605 (BRG, H, US); Upper Rupununi R., near Dadanawa, de la Cruz 2056 (NY, US). Suriname: Tafelberg, Maguire 24259L (H, NY, US); Zanderij, Florschütz 772 (L); Lucie R. to Wilhelmina Mts., Schulz 10353a (L).

Notes: This species is confusingly similar to species of subg. *Cladina*, in particular *C. confusa*, *C. dendroides* and *C. rotundata*. It shares with these the dense, isotomous branching of its podetia, which may form more or less rounded heads. Also the felty surface of the podetia, rugulose by the protuberant algal cell clusters, is very similar. The safest way to distinguish between these is to analyse the chemistry: *C. signata* contains fumarprotocetraric acid with larger or smaller quantities of homosekikaic acid agg. (P+ red, K-, KC-); *C. confusa* lacks fumarprotocetraric acid, but has usnic and perlatolic acid (P-, K-, KC+ yellow); and *C. dendroides* and *C. rotundata* have fumarprotocetraric acid with atranorin (P+ red, K+ yellow, KC-). In the field *C. signata* can usually be recognized by its tendency to form wider patches with rounded surface, while the other species form more separated, semiglobose heads. *C. signata* is strongly browned on light spots, while the other species remain yellowish grey or pale grey. Shade forms of *C. signata* can be very similar in colour to *C. confusa*. With some luck a few squamules can be found in populations of *C. signata* in the field, demonstrating its affinity outside subgenus *Cladina*; in herbarium samples these tend to be absent.

An atranorin-containing strain is known from 4 specimens only, which do not seem to deviate significantly in distribution and ecology.

The closest relatives are probably *C. pulviniformis* and *C. variegata*. They have thicker branches and a different chemistry.

32. **Cladonia sipmanii** Ahti, Fl. Neotrop. Monogr. 78: 304. 2000. Type: Guyana, Region Demerara-Mahaica, Linden Highway, km 7 from Soesdyke, by Housener Farm, savanna patches on white sand, common, 10 m alt., Ahti 52929 (holotype BRG, isotypes B, BM, DUKE, H, MERF, NY, SP, UPS, US).   – Fig. 43

Cladonia salzmannii f. *ascypha* Abbayes, Kew Bull. 11: 263. 1956. Type: Guyana, NW District, Amakura R., 1923, de la Cruz 3547 (lectotype REN, designated by Ahti 2000, isolectotypes BM, NY, US).

Primary thallus evanescent, consisting of rather large, whitish-grey, narrowly laciniate squamules up to 4 mm long necrotic part and hypothallus faintly orange. Podetia 5-15 cm tall, 0.5-1(-2) mm thick, slender to fairly robust, often of indeterminate growth, pale greenish- to brownish grey, not variegated, surface of necrotic bases slightly darkening and often with cyanescent spots; moderately branchy; branching type isotomous to anisotomous dichotomy or trichotomy, more anisotomous in podetia bearing conidiomata or ascomata, then with distinct main stems and short, deflexed side branchlets; axils regularly perforated but usually closed near the tips and frequently also lower down, perforations small, occasionally with more widely gaping funnels; tips erect or slightly divaricate, on fertile podetia often deflexed, furcate. Podetial surface smooth, continuous in upper parts, areolate in lower parts; often slightly squamulose. Podetial wall 240-275 μm; cortex 25 μm; medulla

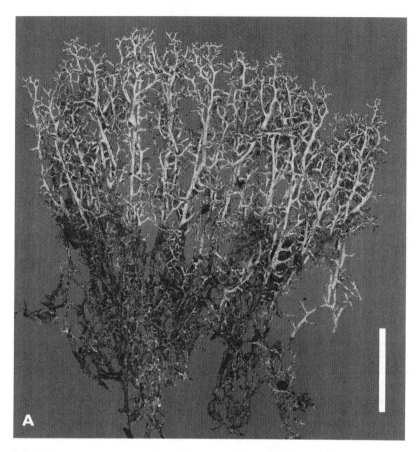

75-100 μm; stereome 125-150 μm, sharply delimited; central canal grooved. Conidiomata cylindrical, not constricted at base, 200-300 × 100 μm; containing red slime. Hymenial discs infrequent, formed at tips of slightly swollen, anisotomously branching podetia; pale brown; 0.3-0.4 mm wide; usually not fusing to form compound discs. Chemistry: thamnolic and barbatic acids, sometimes with 4-O-demethylbarbatic acid and other, unidentified minor compounds (TLC of 30 specimens). One specimen (DePriest 9260) contains squamatic instead of thamnolic acid. Colour reactions: P+ yellow, K+ yellow, KC-, rarely P-, K-, KC-.

Distribution and ecology: An endemic from the Guianas with a single collection from the Venezuelan part of the Guayana Highlands. In the Guianas it is a rather common, widespread species on open, sandy places in savannas, both on the coastal white sand savannas and in the interior of Guyana and Suriname, in French Guianas only in the interior, from 10 to 1000 m alt. Over 60 collections studied (GU: 62; SU: 2; FG: 3).

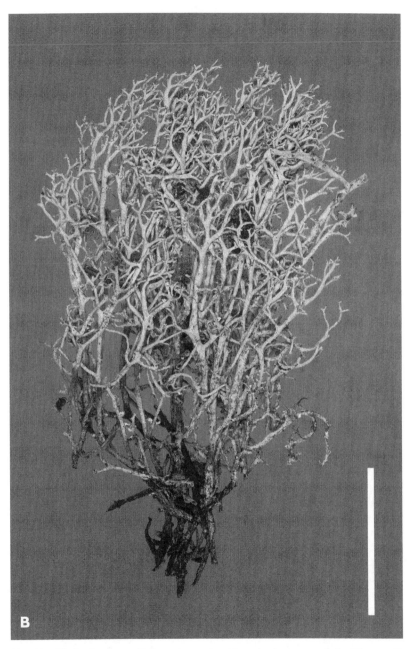

Fig. 43. *Cladonia sipmanii;* A. mature, strongly anisotomous podetia (Sipman & Aptroot 19152 (B)); B. juvenile, mainly dichotomic podetia (Sipman & Aptroot 19501 (B)). Bar = 2 cm.

Selected specimens: Upper Rupununi R., Dadanawa, de la Cruz 2057 (NY, US); Plantation Vrijheid, Linder 967 (FH, TUR-V 15281). Suriname: Zanderij I, Lanjouw & Lindeman 127 (L); Via secta ab Wiawia Bank ad Grote Zwiebelzwamp, savanne bij km 12.4, Lanjouw & Lindeman 874 (L). French Guiana: Inselberg des Montagnes de la Trinité, Jacquemin 2619 (B, L); Tumac Humac, savane roche sur dôme granitique à 1,5 km au nord du Mitaraka Nord, de Granville 1176 (L).

Notes: *Cladonia sipmanii* was earlier confused with the SE Brazilian *C. crispatula*, which is similar but more slender, not brownish, without true cortex and the axils are closed. They might be considered as vicariant taxa. It was also confused with the more distantly related *C. carassensis* Vain., and a record of *Cladonia salzmannii* Nyl. fo. *ascypha* Abbayes (Des Abbayes 1956: 263) probably also belongs here.

Among the Guianas species the closest is *C. hians*. Typical specimens look very different, but more ambiguous specimens occur as well. For the distinction see under that species. *C. huberi*, a rare species from the Venezuelan part of the Guayana Highlands, is also fairly similar in general habit. Its clearest difference is in the fine, blackish tips of the branchlets, (0.5-)1-2 mm long and 0.1-0.2 mm thick.

33. **Cladonia spinea** Ahti, Ann. Bot. Fenn. 23: 215. 1986. Type: Venezuela, Amazonas, Depto. Atabapo, Caño Caname, 150 m alt., Tillett *et al.* 795-130 (holotype H, isotypes MYF, US, VEN).

– Fig. 44

Primary thallus not seen. Podetia 4-8 cm tall, of indeterminate growth, greenish-yellow, forming densely branched cushions, which may form continuous colonies with broad (4-6 cm diam.), rounded heads; branching type isotomous dichotomy (rarely trichotomy), but occasionally parts of thalli have anisotomous tendencies so that main stems may be differentiated, especially towards the base; axils closed, rarely perforated; branchlets 0.2-1.0(-2.0) mm wide, divergent; extreme tips acute, straight, giving a spiny appearance to the lichen, dark brown. Podetial surface in upper parts with fairly compact, corticoid, slightly verruculose layer with minute arachnoid interspaces, towards the base with largely bare stereome with scattered, low areolae containing algal glomerules; squamules absent. Podetial wall 90-175 μm thick, without true cortex but with 5-7 μm thick corticoid outer layer; medulla (incl. algal layer) 30-100 μm; stereome 60-100 μm, horny, fragile; surface of central canal glossy, minutely rugulose. Conidiomata abundant, ovoid to dolioliform, 200-500 × 160-200 μm; containing red slime. Hymenial discs produced at tips of slightly swollen podetia, peltate, pale brown, up to 0.5 mm wide. Chemistry: usnic acid with thamnolic acid or barbatic

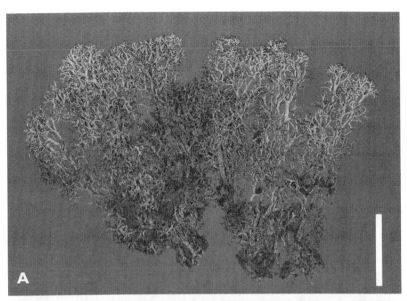

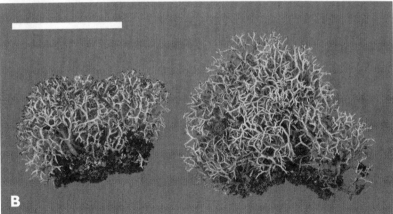

Fig. 44. *Cladonia spinea*; A. luxuriant form (Sipman 40348 (B)); B. stunted form (Sipman 57156 (B)); Bar = 2 cm.

acid or with additionally squamatic acid, rarely with usnic acid alone (TLC of 28 specimens). Colour reactions: P- or P+ yellow-orange, K- or K+ yellow-orange, KC+ yellow.

Distribution and ecology: An endemic of the Guianan Shield and surroundings, being mainly found in Venezuela, Colombia, Brazil and Guyana, at low elevations (10-500 m alt., up to 1200 m alt. in the Gran Sabana, Venezuela). In the Guianas so far found only in Guyana, mainly on the sandstone tableland of the Pakaraima Mts., but also in white sand savannas near the coast. It grows terrestrial in open places in scrub savanna, on sandstone flats or on sand, from 10 to 1200 m alt. Over 100 collections studied (GU: 61).

Selected specimens: Exsicc.: Lichenotheca Latinoamericana 115. Guyana: Region Potaro-Siparuni, Kaieteur Falls National Park, DePriest 9254 (H, US); Region Cuyuni-Mazaruni, Chi-Chi Mountain range, Pipoly 10200 (H); Demerara-Mahaica Region, Linden Hwy, DePriest 9227 (H, US); East Demerara Distr., Timehri, Sipman & Aptroot 19498 (H, L); Hoosookea Savanna, Jenman 3711 (BM, BRG, NY); Upper Mazaruni Distr., Makwaima savanna near Mayaropai, Sipman & Aptroot 18537 (B); Essequibo R., Kurupukari, Smith 2177 (BM, G, U, US); Pakaraima Mts., Mt. Aymatoi, Maas et al. 5703 (H, L).

Notes: *Cladonia spinea* is a conspicuous lichen, easily recognized by its broad, spiny, bright yellowish cushions. It shares this aspect with a number of related species, *C. steyermarkii*, *C. substellata*, *C. vareschii*, and to a lesser extent *C. subreticulata* and *C. sufflata*. While *C. subreticulata* is easily recognized by its very thick, often over 2 mm wide, irregularly branched podetia and *C. sufflata* by its scarce branchings and soft stereome, the others are more difficult to tell apart. *C. spinea* is distinguished from all by the combination of a glossy inner surface of the horny stereome, the scarcety of main stems, and the thin podetium walls, less then 170 μm thick. Yellowish specimens of *C. peltastica*, containing usnic acid, may also form dense cushions, but they have more slender podetia, under 0.5 mm wide, with longer internodes, they may bear thallus squamules, and the hymenial discs develop on much swollen podetia. Occasionally, scarcely branched specimens with distinct main stems resemble *C. flavocrispata*, see note under that species.
A strain with usnic and decarboxythamnolic acid was found outside the Guianas.

34. **Cladonia sprucei** Ahti, Ann. Bot. Soc. Zool. Bot. Fenn. Vanamo 32(1): 100. 1961. – *Cladina sprucei* (Ahti) Ahti, Beih. Nova Hedwigia 79: 40. 1984. Type: Brazil, Amazonas, Manaus, Umirisál opposite to the mouth of Rio Negro, Spruce, Lich. Amaz. And. 17 (holotype H-NYL 37625, isotypes BM, G, NY, PC). – Fig. 45

Primary thallus absent. Podetia 7-12 cm tall, of indeterminate growth, pale grey, at the extreme apical tips browned and, especially downwards, sometimes blackened, dead basal parts finally clearly melanotic, with mature internodes 2-3 mm long, at the tops of separate podetia 1-2 cm wide, forming dense but not extremely richly branched, flat-topped colonies;

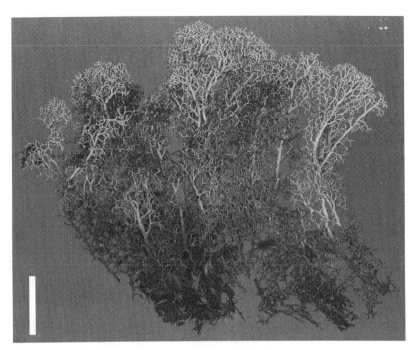

Fig. 45. *Cladonia sprucei* (Sipman 40290 (B)). Bar = 2 cm.

branching type anisotomous dichotomy but subisotomous tendencies distinct, trichotomy very scarce; axils closed or with small holes; ultimate apical branchlets erect to divaricate, attenuate or blunt; main stems rather distinct in lower parts, 0.8-1.6 mm thick. Podetial surface arachnoid and smooth near the tips, arachnoid and distinctly verruculose towards the base. Podetial wall in main stems c. 170-250 μm thick; stereome c. 70-100 μm. Conidiomata containing red slime. Hymenial discs not seen, probably brown, globular, on terminal branchlets. Chemistry: atranorin and fumarprotocetraric acid as major substances, sometimes also psoromic acid; additionally with traces of protocetraric and confumarprotocetraric acids, and occasionally ursolic acid and 2-3 unknown substances (TLC of 20 specimens). Colour reactions: P+ fast red, K+ yellow, KC-.

Distribution and ecology: A neotropical endemic known from the Amazonian lowlands in Colombia, Venezuela, the Guianas and northern Brazil, and extending to mid elevations in the Guayana Highlands. In the Guianas it is the commonest member of subgenus *Cladina*, but nevertheless only found in Guyana, and there mainly in the Pakaraima Mts. It grows especially on open places in scrub savanna on the white sands of alluvial areas and on sandstone plateaux, from 50 to 1000 m alt. Over 100 collections studied (GU: 43).

Selected specimens: Exsicc.: Lichenotheca Latinoamericana 6 (as *Cladina sprucei*). Guyana: Atkinson, St. Cuthbert's Trail, U.G. Bio 106 (NY); Pomeroon R., Jenman 8051a (BM); Upper Mazaruni Distr., Mt. Latipu, Sipman & Aptroot 19129 (B); 5 km N of Waramadan, Sipman & Aptroot 19244 (B); Kamarang-Pwi Pwi Mt., N of Waramadan, Sipman & Aptroot 19387 (B); Region Potaro-Siparuni, Kaieteur Falls National Park, Pipoly 9851 (H); Region U. Demerara-Berbice, Linden-Soesdyke Hwy, Pipoly 9734 (H); Region Demerara-Mahaica, Linden Highway, DePriest 9224, 9225 (H); Region Cuyuni-Mazaruni, Chimoweing village, Pipoly 10507 (H); Region 7 (Upper Mazaruni Distr.), N of Paruima Mission, Aymatoi savanna, Sipman 39876 (B, BRG).

Notes: For differences with its closest relative, *C. atrans*, see there. *C. dendroides* has less distinct main stems and its apical branches are not deflexed in one direction ("combed"), as often in *C. sprucei*.
Another similar species, *C. rangiferina* var. *abbayesii* has been reported from the Guianas but its presence here is unlikely in view of the arctic-alpine/andean general distribution of the species. It can be recognized by its more brownish to violet colour: its tips are browned over a longer distance, and browned branches occur to near the apex of the podetia, in particular in sunny situations, where *C. sprucei* remains grey, unless when deteriorating. A report from Guyana (Ahti 2000) is referred here to *C. sprucei*, because the specimen contains psoromic acid as minor compound, a substance unknown from *C. rangiferina* but locally common in *C. sprucei*.
The chemical strains with and without psoromic acid seem equally common and widespread.

35. **Cladonia steyermarkii** Ahti, Ann. Bot. Fenn. 23: 217. 1986. Type: Venezuela, Bolívar, Dist. Piar, Macizo del Chimantá, Acopán-tepui, 1950 m alt., Ahti *et al.* 45148 (holotype VEN, isotypes B, COL, H, NY, TNS, US). – Fig. 46

Primary thallus not seen. Podetia 3-7 cm tall, 0.8-1.2 mm thick when branching more isotomous, or up to 2(-3) mm when branching strongly anisotomous, of indeterminate growth, pale to intensely yellow, browned at the extreme tips, necrotic bases turning dark to reddish-brown, typically growing erect in dense colonies, usually sparsely branched; branching type irregular, anisotomous dichotomy, more rarely trichotomy or tetrachotomy; main stems tend to be robust, well differentiated, but morphs with thin, more intricately branched podetia and poorly defined main stems also occur; axils closed or perforated (some with widely gaping holes); branchlets mainly short, less than 5 mm long, with erect to slightly recurved, acute tips. Podetial surface smooth, occasionally uneven near

Fig. 46. *Cladonia steyermarkii* (Sipman 39906 (B)). Bar = 2 cm.

the tips, distinctly corticate, matt to somewhat glossy, maculate. Podetial wall fragile, 150-200 μm thick; cortex well developed and often almost as thick as the stereome, 25-30 μm; medulla almost absent or up to 30 μm thick; stereome 100-150 μm, or only 35-40 μm below the immersed, larger medullary algal glomerules; surface of central canal matt, white pulverulent. Conidiomata conical, 100-120 × 80 μm, containing purple slime. Hymenial discs not seen. Chemistry: usnic and squamatic acids (TLC of 4 specimens). Colour reactions: P-, K-, KC-.

Distribution and ecology: A neotropical endemic centered in the Guayana Highlands, ocasionally found in the Venezuelan Andes, common at middle and high elevations (1000-2580 m). In the Guianas all available records are from the Aymatoi savanna in the sandstone tableland of the Pakaraima Mts., close to the Venezuelan border, where the species grows in scrub savanna on rock plateaux which stay wet long due to water trickling from the surrounding vegetation after rain, from c. 1000 to 1200 m alt. Over 20 collections studied (GU: 6).

Selected specimen: Guyana: Region 7 (Upper Mazaruni Distr.), N of Paruima Mission, Aymatoi savanna, Sipman 39870 (B, BRG).

Notes: *Cladonia steyermarkii* is most easily confused with *C. substellata*, which resembles it closely in colour and in the pruinose wall of its central canal. *C. substellata* differs by its thin corticoid outer layer, rather than a thick cortex; its stereome has a very irregular outer surface, and in the herbarium it develops steroid needles after a few years. Also it is a smaller plant with less pronounced main stems. Another fairly similar species in colour and branching pattern, *C. crassiuscula*, differs by its central cylinder, which is not a compact stereome but composed of more loosely conglutinated hyphae. Certain forms are somewhat similar in branching to *C. sufflata*, which can be easily separated by its more greyish colour.

Cladonia steyermarkii is also very close to *C. uncialis* (L.) Weber ex F.H. Wigg., a widespread northern circumpolar, arctic to temperate species. The latest differs by the absence of thamnolic acid (usnic acid alone or with squamatic acid present), it has a better developed medulla, a more continuous algal layer, a glossy cortex, a more irregular branching and rarely as intensely yellow a colour as in *C. steyermarkii*.

Outside the Guianas the dominant chemotype contains usnic thamnolic acids, with traces of decarboxythamnolic acid and the chemotype with squamatic acid is rare. Small amounts of barbatic acid and unknown substances may also occur in both chemotypes.

36. **Cladonia subdelicatula** Vain. ex Asahina, J. Jap. Bot. 38: 1. 1963. Type: Brazil, Rio Grande do Sul, Mun. Venâncio Aires, Faxinal Tamanco, Jürgens 5735 (135) (holotype TUR-V 15098a, isotypes FH, FR, US). – Fig. 47

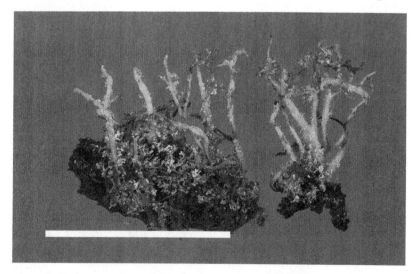

Fig. 47. *Cladonia subdelicatula* (Boom et al. 40089 (hb. v. d. Boom)). Bar = 2 cm.

Primary thallus persistent, consisting of narrowly laciniate squamules, 2-5 mm long, esorediate. Podetia 1.5-3(-7) cm tall, 0.2-0.7 mm thick, of determinate growth, whitish to greenish, simple to slightly branched by dichotomy; axils closed; without axillary funnels; tips narrow, subulate. Podetial surface mostly over a large surface ecorticate and felty, with corticate verruculae and more or less corticate granules, often with abundant, finely laciniate squamules, esorediate but sometimes with soredioid granules. Podetial wall 150-400 μm thick; cortex 20-30 μm; medulla 0-20 μm; stereome 150-340 μm, hard, pellucid; central canal slightly papillate. Conidiomata on basal or podetial squamules or at tips of podetia; cylindrical, 400 × 200 μm, usually not constricted at base; conidial slime not seen. Hymenial discs infrequent, 0.5 mm wide, light brown. Chemistry: thamnolic acid, occasionally also barbatic acid, and traces of decarboxythamnolic acid and unknown compounds (Ahti 2000). Colour reactions: P+ yellow, K+ yellow, KC-.

Distribution and ecology: A widespread species in the Neotropics, where it was previously known from Amazonia, SE Brazil and Paraguay; also recorded in the Palaeotropics (La Réunion). It grows on rotten wood, tree bases and rocks in shaded habitats, especially restinga woodlands, but extending to cloud forests. In the Guianas, a few records exist from the Pakaraima Mts. in Guyana, on mossy bark of living trees, from 400 to 500 m alt. Probably the species is much more widespread than estimated based on recorded collections. Because it is inconspicuous and easily confused with other species, both sampling and correct identification are hampered. Over 100 collections studied (GU: 3; SU: 1).

Selected specimens: Guyana: Potaro-Siparuni Region, 1 km SW of Paramakatoi, Ahti *et al.* 53333 (B, BRG, H, US); id., Kaieteur Falls National Park, Ahti *et al.* 53054 (H); Cuyuni-Mazaruni Region, Mt. Aymatoi, DePriest *et al.* 10418 (BRG, DUKE, H, NY, US, VEN).

Notes: A rather variable and easily mistaken species. In particular, the degree of branching and the production of soredioid granules and squamules are very variable, and accordingly, the species shows a resemblance to quite different lichens. Scarcely branched plants with few propagules may resemble *C. didyma*, but differ because the stereome remains covered by a whitish, felty layer, and does not become brownish-translucent.
Scarcely branched, but strongly granulose plants resemble *C. subradiata*, which may occur in the same habitat. They can be distinguished by their chemistry.
Richly branched, squamulose plants come close to *C. persphacelata* and *C. polystomata*. The first has a smooth, not felty surface on the podetia and grows on rock, while the second has conspicuous, wide open axils and growns preferably on sand.

37. **Cladonia subradiata** (Vain.) Sandst., Abh. Naturwiss. Ver. Bremen
25: 230. 1922. – *Cladonia fimbriata* [var.] *chondroidea* [subvar.]
subradiata Vain., Acta Soc. Fauna Fl. Fenn. 10: 338. 1894. Type:
Brazil, Minas Gerais, Caraça, 1400 m alt., 1885, Vainio s.n. (lectotype
TUR-V 19517, designated by Ahti 1993). – Fig. 48

Primary thallus persistent to evanescent, consisting of 2-4 mm long and
1-2 μm wide, crenulate to laciniate, thin and soft squamules, esorediate
or granular-sorediate below and at margins. Podetia abundant, 0.8-
2.5 cm tall in mature state, slender, 0.4-1.0 mm thick, of determinate
growth, whitish-grey, not blackening with age, unbranched or sparsely
branched, gradually tapering towards the acute or bluntish tips before
scyphus formation; scyphi absent in very young podetia, but narrow
scyphoid tips may be visible at an early stage, and scyphi are frequent
in mature podetia, (0.6-)1.2-2(-4) mm wide, shallow, marginally (very
rarely centrally) proliferating, proliferations usually soon terminated
by hymenial discs; occasionally the scyphi may be laterally completely
flattened. Podetial surface slightly corticate at the very base, mostly
covered by isidioid microsquamules (0.1-0.2 mm long) and granules,
and with larger squamules towards the base, in upper parts and in
deep shade so finely granulose as to be called clearly sorediate, but the
soredial granules often elongate in shape. Podetial wall 180-200 μm

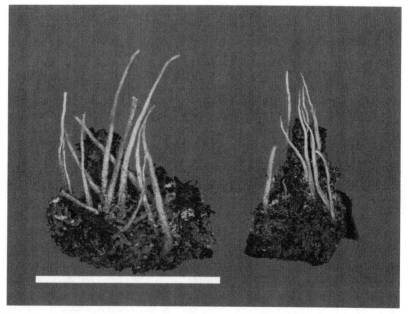

Fig. 48. *Cladonia subradiata* (Ahti et al. 52913 (B)). Bar = 2 cm.

thick; cortex 0-40 μm, granulose layer 0-50 μm; medulla 15-25 μm; stereome fairly hard, 100-150 μm, inner surface minutely papillose and striate. Conidiomata common, either basal on young squamules, sessile or stalked, or terminal at podetial tips or on young scyphi, 0.1-0.3 × 0.-0.2 mm, sessile to shortly pedicellate, bell-shaped to irregularly pyriform, strongly constricted at base; containing hyaline slime. Hymenial discs fairly common, dark brown, primordia characteristically pale brown, in botryose pedicellate groups on scyphal margins or terminal on subulate podetia, single discs rarely more than 2-3 mm wide. Chemistry: fumarprotocetraric acid with traces of protocetraric, quaesitic (subconstant) and confumarprotocetraric acids (TLC of 29 specimens). Colour reactions: P+ orange red, K-, KC-.

Distribution and ecology: Widespread in tropical areas in America, Asia, Australasia and Africa, and extending into temperate areas in eastern N America. In the Neotropics it is widespread throughout tropical America, from lowland rain forests to subarid cerrados and up to 4200 m alt. in the Andes. The primary habitat is rotten wood in somewhat shady forest, but it also grows on humous and sandy banks, rock outcrops, termite hills and other decaying organic material, often in man-made habitats. Probably widespread in the Guianas from 10 to 1200 m alt. Over 500 collections studied (GU: 33; SU: 16; FG: 16).

Selected specimens: Guyana: Amakura R., de la Cruz 3514 (NY, US); Mt. Iramaikpang, Smith 3654 (NY). Suriname: Casipoerakreek, Benjamins s.n. (L); Maratakka, Florschütz 2030 (L); Zanderij, Florschütz 656 (L). French Guiana: Cayenne, Broadway 842 (NY, US); St-Elix, Sastre 513 (H).

Notes: *Cladonia subradiata* is a difficult species to recognize, because it is often found as stunted plants. It bears a considerable resemblance to the holarctic *C. coniocraea* (Flörke) Spreng., which has ordinary farinose soredia and a distinct white medulla. Among the sympatric species in the Guianas, it is particularly similar to *C. polyscypha*, and the possibility cannot be excluded that some of the material attributed here to that species actually belongs to *C. subradiata* (see note under *C. polyscypha*). The soredium-like granules are often clumped together in minute, squamule-like structures, not known from other species.

38. **Cladonia subreticulata** Ahti, Ann. Bot. Fenn. 10: 168. 1973. Type: Brazil, Minas Gerais, Caraça, 1885, Vainio s.n. and in Sandstede: Clad. Exs. 1194 (holotype TUR-V 13546, isotypes BM, C, FH, G, H, MIN, TUR-V 13552, UPS, US). — Fig. 49

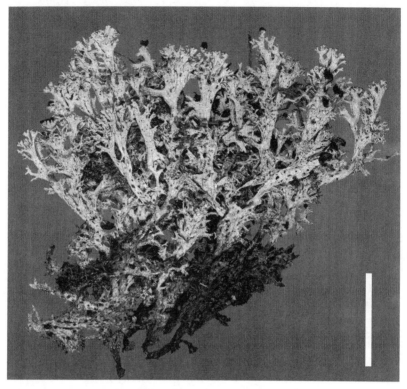

Fig. 49. *Cladonia subreticulata* (Sipman 39872 (B)). Bar = 2 cm.

Primary thallus unknown. Podetia 3-10 cm tall, main stems 2-5 (-10) mm wide, width highly variable because of the strong and irregular branching, of indeterminate growth, pale yellowish grey, occasionally with lilac patches in the morbid basal parts, which otherwise turn grey, slightly browned at the extreme apical tips, stout and blunt, usually somewhat expanding towards the apices and at the axils, then forming very irregular scyphi-like structures with numerous short branchlets at the margins, forming densely packed cushions; branching pattern irregular anisotomous polytomy, with tetrachotomy predominant; axils at first closed, soon perforated, and with age also the podetial walls become perforated, split and finally fibrose at base. Podetial surface matt, slightly arachnoid to fairly compact, ecorticate. Podetial wall 150-200 μm thick, mostly filled with cartilaginous, longitudinal stereome strands of various sizes, which do not form a distinct central cylinder, but almost reach to the outer surface in places; surface layer 10-30 μm, consisting of closely interwoven, 6-7 μm thick hyphae; algal layer discontinuous; medulla very poorly developed, irregular; central canal with a tomentose to pruinose surface, not or indistinctly reticulate. Conidiomata terminal, not seen in mature condition. Hymenial discs very numerous on short stalks at podetial ends, up to 0.3 mm wide, pale brown. Chemistry: usnic acid usually with fumarprotocetraric acid (TLC of 11 specimens). Steroid crystals may develop at the tips in the herbarium. Colour reactions: P- or P+ orange red, K-, KC+ yellow.

Distribution and ecology: A neotropical endemic known form the Guayana Highland in Venezuela and Guyana, the E-side of the Andes in Peru and Ecuador and the southern Brazilian Highlands. In the Guianas rather widespread on the sandstone tableland of the Pakaraima Mts. in Guyana. It grows on open places in scrub savanna, on sandstone plateaux, from 500 to1200 m alt., and may be favoured by fire. Over 100 collections studied (GU: 17).

Selected specimens: Guyana: Region Cuyuni-Mazaruni, 2 km W of Chi-Chi Falls, Pipoly *et al.* 10203 (H); Pakaraima Mts., Mt. Aymatoi, Maas *et al.* 5722 (B, H, L); Cuyuni-Mazaruni Region, 8.6 km NE of Imbaidamai, Hoffman 1696 (H, US); Upper Mazaruni Distr., Mt. Latipu N of Kamarang, Sipman & Aptroot 19155 (B, H); Karowtipu Mtn., Boom & Gopaul 7613 (BRG, NY).

Notes: *Cladonia subreticulata* is one of the most conspicuous, and an easily recognizable *Cladonia* species, because of its unusually thick, irregularly branched, perforated podetia, which may form large, conspicuous cushions. The perforated podetia may remind of the genus *Cladia*, in which the podetia are greenish to brownish, not pale yellowish, and glossy, not matt.
Outside the Guianas an additional chemotype occurs with (-)-usnic acid (sometimes with isousnic acid) plus the stictic acid complex, with stictic, constictic, cryptostictic (±), norstictic (±), and connorstictic (±) acids - all these apparently highly concentrated at podetial tips; one or more unknown terpenoids (including zeorin), as well as fatty acids and other unknowns may also occur in trace amounts. Steroid crystals may develop at the tips in the herbarium.
The record of *C. reticulata* (J.L. Russell) Vain. (Sandstede 1932) most probably belongs here.

39. Cladonia subsphacelata Sipman & Ahti, Phytotaxa 93: 1. 2013. Type: French Guiana, Savane Roche de Virginie, Bassin de l'Approuague, 120 m alt., Fourré isolé de savane roche, 12 Feb. 1991, G. Cremers & P. Petronelli 11900 (holotype B).   – Fig. 50

Primary thallus persistent, consisting of up to 0.5 cm long squamules which are deeply divided into c. 0.5 mm wide, elongate laciniae, attenuated and often almost stalk-like at the base, with rather smooth to corticoid surface towards the base and sometimes with ochraceous streak. Podetia up to 3 cm tall and 0.3-0.8 mm thick, of determinate growth, grey to usually more or less brown, at base almost black, somewhat branched; branching type irregular anisotomous dichotomy, rarely trichotomy or tetrachotomy; axils closed or with usually small openings; tips often divided into 2-10 short branchlets; central canal very thin, c. 0.1 mm wide. Podetial surface smooth and mostly corticate, not shiny, finally being rather densely squamulose, smooth inbetween, esorediate; mature squamules narrow, laciniate and imbricate, up to 4 mm long, pointing downward but with recurved tips, often glossy. Podetial wall 200-300 μm thick; cortex (0-)25-40 μm; medulla thin, 10-25 μm (including the algae); stereome distinctly delimited, horny, thick, 200-250 μm, inner surface glossy. Conidiomata not seen. Hymenial discs not seen. Chemistry: squamatic and didymic acids (TLC of 4 specimens). Colour reactions: P-, K-, KC-; UV+white.

Fig. 50. *Cladonia subsphacelata;* type (Cremers & P. Petronelli 11900 (B)). Bar = 2 cm.

Distribution and ecology: Species known so far only from French Guiana and Guyana, found in sananna over rock from 100 to 400 m alt. Five collections studied (GU: 3; FG: 2).

Selected specimens (paratypes): Guyana: Potaro-Siparuni Region, Kaieteur Falls National Park, around the airstrip, Sipman 40326 (B, BRG); Ahti *et al.* 53048a (B, H, US); trail to Johnson's View, DePriest 9375 (H). French Guiana: Savane Roche de Virginie, Bassin de l'Approuague, Cremers & Petronelli 11899 (B).

Notes: For differences with related species see under *C. persphacelata*.

40. **Cladonia subsquamosa** Kremp. in Warming, Vidensk. Meddel. Naturhist. Foren. Kjøbenhavn 5: 366. 1874 ('1873'). Type: Brazil, Rio de Janeiro, "Serra d'Estrella & Petrópolis", Warming 233 (lectotype C, designated by Ahti 1993, isolectotypes G, M, TUR-V 19478, UPS). – Fig. 51.

Primary thallus often persistent, consisting of 1-3 mm long and 1-2 mm wide, soft and fragile, laciniate, often imbricate squamules, upper side more or less convex, lower side fluffy, esorediate but often granular or secondarily sorediate. Podetia 1-1.5 cm tall, stalks 0.6-1 mm wide, of determinate growth, greyish green, usually not blackening at base, denuded surface easily browned, especially at tips, always forming scyphi, even in young stage, rarely (in deep shade) producing subulate branchlets; scyphi 2-3 (-5) mm wide, broadest at margins; margin thin, erect or slightly recurved,

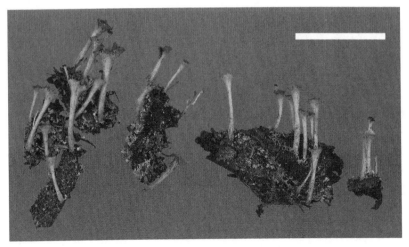

Fig. 51. *Cladonia subsquamosa* (Sipman 39838 (B)). Bar = 2 cm.

soon dentate and sometimes repeatedly (1-3 times) producing 1-6 marginal, scyphi-like (rarely subulate) proliferations. Podetial surface not corticate at all or slightly so at the very base, mostly coarsely or finely sorediate; soredia often in part mixed with minute (0.1-0.2 mm long), isidioid phyllidia like in *C. subradiata*, in part with globose, corticate granules or microsquamules, in old podetia sometimes densely beset with such structures; soredial layer thin and loosely attached, soon in part easily disintegrating, leaving the white or brownish, grooved stereome bare; occasionally totally microsquamulose (like the type specimen, hence the name) or also macrosquamulose; scyphi, when viewed from above, appear thinly and loosely granular sorediate. Podetial wall 340-380 μm, sorediallayer up to 80 μm; medulla extremely thin, scarcely forming a white layer, up to 40 μm; stereome hard and thick, glassy rather than opaque, 240-280 μm; inner surface minutely papillose; scyphi wall usually translucent (best seen in fresh specimens in moist condition!). Conidiomata terminal, semiglobose to pyriform, little constricted at base; containing hyaline slime. Hymenial discs rather common, 1-5 mm wide, light (often) to dark brown, on long (4-7 mm) stalks on the scyphal margins. Chemistry: fumarprotocetraric acid with trace amounts of protocetraric, quaesitic (constant), confumarprotocetraric acids (TLC of 30 specimens) or other, mostly unknown minor substances (TLC of 2 specimens). Colour reactions: P+ orange red, K-, KC-.

Distribution and ecology: A common pantropical species, widespread throughout the Neotropical area, where it is one of the most frequent species from the lowlands up to 4100 m alt. in the Andes, primarily on rotten wood and tree bases in semi-open to shaded, often disturbed habitats, but also growing on soil. Rather infrequently found in the Guianas, so far only in Guyana, where it was growing on termite mounds or terrestrial on open places, from 10 to 1000 m alt. Over 500 collections studied (GU: 9).

Selected specimens: Guyana: Region Potaro-Siparuni, Kaieteur Falls National Park, Ahti 53015 (H); Region U. Demerara-Berbice, Linden-Soesdyke Hwy, Pipoly 9752 (H); Upper Mazaruni Distr., Makwaima savanna, Sipman & Aptroot 18538 (B).

Notes: This species is easily recognizable among the "cup lichens" of the Guianas by its grey rather than greenish colour, caused by the absence of usnic acid. It is finely sorediate, like *C. mollis*, but usually with longer stalks and shorter cups.
Cladonia subsquamosa Kremp. should not be confused with *C. subsquamosa* (Nyl. ex Leight.) Cromb., its much used later homonym, a *nom. illeg.*, which is a very different taxon often included as a chemotype in *C. squamosa*.
The record of *Cladonia pyxidata* (L.) Fr. var. *chlorophaea* "Flörke" (Vainio 1894) most probably belongs here.

41. **Cladonia substellata** Vain., Acta Soc. Fauna Fl. Fenn. 4: 271. 1887.
Type: Brazil, Minas Gerais, Caraça, 1400 m alt., 1885, Vainio s.n.
(lectotype TUR-V 13632, designated by Ahti 1973, isolectotype
H-NYL 37737).   – Fig. 52

Primary thallus not seen in Guianas specimens, acc. to Ahti (2000)
consisting of thick, 1-2 mm wide squamules with pulverulent lower
side and dentate margins. Podetia 3-7 cm tall, 0.5-1.5(-2) mm thick,
slender, of indeterminate growth, pale yellow to greyish green, very
slightly browned at extreme tips, necrotic base pale grey, forming dense
colonies, much branched; branching pattern regular, often anisotomous
dichotomy; axils closed or open, never forming scyphi. Podetial surface
matt, smooth, often with fur of fine hair-like steroid crystals near the tips
after a few years in the herbarium. Podetial wall 140-225 μm thick, surface
layer corticoid, thin, consisting of 8-10 μm thick hyphae; medulla, incl.
discontinuous algal layer, 10-70 μm thick, rather irregularly delimited
against the fibrose stereome, which is forming most of the podetial
wall; central canal furrowed, pruinose. Conidiomata terminal, conical
to ovoid-cylindrical, 150-200 × 150 μm, not or little constricted at base,
containing hyaline slime; conidia small, 4-6 × 0.5 μm. Hymenial discs
infrequent, at tips of apically much branched, slightly thicker podetia,
pale brown. Chemistry: usnic acid, sometimes with fumarprotocetraric
acid (TLC of 9 specimens). Steroid crystals are produced in abundance
at the podetial tips in the herbarium. Colour reactions: P+ slowly yellow
or P-, K-, KC+ yellow.

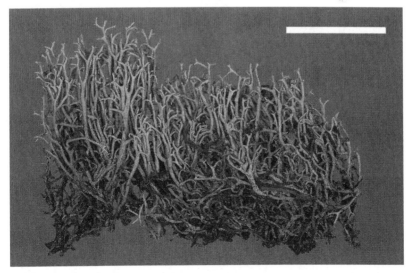

Fig. 52. *Cladonia substellata* (Sipman 40334 (B)). Bar = 2 cm.

Distribution and ecology: A neotropical species known from the Guayana Highland (Cerro de la Neblina, Cerro Duida), NE Brazil, the SE Brazilian highlands, Uruguay and Paraguay. It occurs on sand in coastal restinga, on rocks, but also rocky soil, uncommon, but may be locally very abundant, up to 1400 m alt. In the Guianas known so far only from Guyana, mainly in the Pakaraima Mts., but also collected once near Linden. It grows in clearings in shrub savanna, on sand or sandstone flats, from 50 to 1200 m alt. Over 50 collections studied (GU: 12).

Selected specimens: Exsicc.: Lichenotheca Latinoamericana 117. Guyana: Demerara-Mahaica Region, Linden Hwy, DePriest 9227 (H, US); Region Potaro-Siparuni, Kaieteur Falls National Park, Ahti 53042 (H); Upper Mazaruni Distr., 8 km N of Kamarang, Mt. Latipu, Sipman & Aptroot 19160 (B); Region 7 (Upper Mazaruni Distr.), N of Paruima Mission, Aymatoi savanna, Sipman 39896 (B, BRG).

Notes: *Cladonia substellata* is a very distinct S American species. It is particularly characterized by the fibrose stereome, the steroid crystals (visible with stereomicroscope at high magnification) and the stictic acid complex (outside the Guianas). It has three very closely related species, viz. *C. stenroosiae* Ahti in Brazil, *C. dimorphoclada* Robbins and *C. labradorica* Ahti & Brodo in N America, which all contain usnic acid only, and are confined to sandy habitats.

The Guianan material differs slightly from the Brazilian populations of *C. substellata* because the outer side of the stereome seems less irregular. The usnic and fumarprotocetraric acid chemotype is known only from Venezuela and Guyana. A further chemotype exists with usnic, stictic, constictic, cryptostictic, norstictic (inconstant), and connorstictic (inconstant) acids. The stictic acid complex is often so scarce that the colour reagents P and K may fail to give positive reactions.

42. **Cladonia sufflata** Ahti, Lichenologist 22: 265. 1990. Type: Venezuela, Bolívar, Dist. Piar, Macizo del Chimantá, Acopán-tepui, Ahti *et al.* 45078 (holotype VEN, isotypes B, DUKE, H, NY, US).

– Fig. 53

Primary thallus evanescent, not seen. Podetia 5-7 cm tall, 1-3 mm thick, of indeterminate growth, robust, pale greenish- or greyish yellow, lower down variegated yellowish and brownish, somewhat branched, forming open colonies, often reminiscent of inflated, thin fingers; branching pattern anisotomous polytomy or dichotomy, most branchlets very short (less than 0.5 cm) and main stems very dominant from the tips; axils usually perforate, often gaping, also lateral perforations present; tips acute

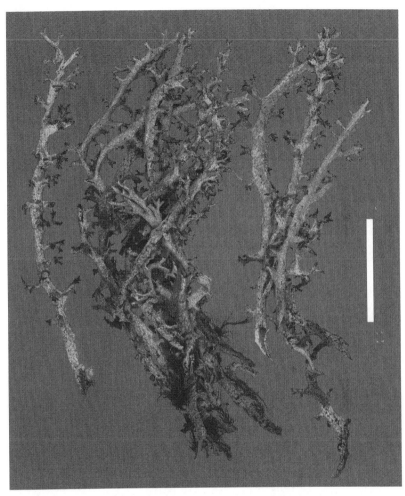

Fig. 53. *Cladonia sufflata* (Sipman & Aptroot 19161 (B)). Bar = 2 cm.

to almost blunt. Podetial surface matt, smooth near the tips, ecorticate, continuously tomentulate, becoming shallowly areolate towards the base, areolae hardly tomentulate but interspaces revealing the browned, felty medulla; esquamose. Podetial wall 150-240 μm thick; cortex lacking or (on algal glomerules) up to 30 μm thick, hyphae clearly encrusted with crystals; medulla 30-40 μm; stereome 90-100 μm, distinct but soft in structure; surface of central canal felty. Conidiomata not seen. Hymenial discs not seen. Chemistry: usnic and thamnolic acids with accessory decarboxythamnolic acids (TLC of 3 specimens). Colour reactions: P+ yellow, K+ yellow, KC+ yellow or KC-.

Distribution and ecology: The species seems restricted to the Guayana Highland and is common in adjacent Venezuela, in open mountain bogs or gravelly creeksides, where it can be locally abundant. In the Guianas it is occasionally found in the Pakaraima Mts. on the sandstone tablelands. Here it seems restricted to the higher, more humid parts, and it grows on thin soil covers often soaked with rainwater, from 1000 to1200 m alt. Over 25 collections studied (GU: 5).

Selected specimens: Guyana: Upper Mazaruni Distr., Mt. Latipu, 8 km N of Kamarang, Sipman & Aptroot 19161 (B); Region 7 (Upper Mazaruni Distr.), N of Paruima Mission, Aymatoi savanna, Sipman 39931 (B, BRG).

Notes: *Cladonia sufflata* is easily recognized by the thick, pale greyish yellow podetia, which are variegated towards the base. Its stereome is soft, and its podetia are predominantly anisotomously branched, with main stems developed within 1-2 mm from the tips. It is most similar to *C. steyermarkii* and *C. vareschii*. The first has well developed, thick cortex and its podetia are usually not over 1.2 mm thick. The second has a thick and hard stereome, a central canal with glossy surface, and its podetia are usually not over 1.5 mm thick.
An usnic acid-deficient, ash-grey chemotype occurs in the Venezuelan part of the Guayana Highlands.

43. **Cladonia termitarum** Ahti, Phytotaxa 93: 1. 2013. Type: Guyana, Potaro-Siparuni Region, Kaieteur Falls National Park, S side of airstrip, 0.5° 10'N, 59° 29'W, 400 m alt., on termite structure on premontane sclerophyll forest floor, 1996, T. Ahti *et al.* 53023 (holotype BRG, isotypes B, H, US).   – Fig. 54

Primary thallus persistent, appearing crustose but consisting of 0.3-1 mm long, 0.1-0.5 mm wide, pale greenish to brownish grey squamules, usually very short and irregular, convex, thick, often warty or almost cylindrical, often densely aggregated, crenulate at ends and margins, with eroded, granulose patches extending on the underside of the squamules; necrotic basal parts becoming orange. Podetia often numerous, very short (0.5-2 mm), of determinate growth, greenish grey; stalk smooth (little corticate), granulose or densely squamulose. Podetial surface areolate-verruculose. Podetial wall not measured. Conidiomata scattered, subspherical, black, 0.2 mm tall, after ejaculation of conidia ostiolum gaping open with black, toothed margins and red slime visible

Fig. 54. *Cladonia termitarum;* type (Ahti *et al*. 53023 (B)). Bar = 1 cm.

inside. Hymenial discs sparsely produced, on very short (0.5-2 mm), smooth, granulose or squamulose podetia, flat, purple. Chemistry: PD+ fast brick red, containing the fumarprotocetraric acid complex; the purple pigment rhodocladonic acid in hymenial discs and conidiomata.

Distribution and ecology: A Guianas endemic, known so far only from three collections made near the Kaieteur Falls, Guyana. It was exclusively observed on old forest floor termite structures, at c. 400 m alt. These structures are very common in the Amazonian and associated lowland rainforests, therefore the species is expected to be more widespread. (GU: 3).

Specimens examined (paratypes): Guyana: Potaro-Siparuni Region, type locality, termite mound, Ahti 53020 (BRG, H, US); Kaieteur Falls National Park, N side of airstrip, termite mound, Ahti 53035 (B, BRG, H).

Notes: The short-stalked, bright red ascomata in combination with the presence of fumarprotocetraric acid set this species well apart from all Guianas lichens. The only similar species is *Cladonia ahtii* Stenroos, known from SE Brazil, which belongs to the group of *C. miniata*, and differs by the shape of the squamules, thick and rounded, with a flat to concave upper surface.

44. **Cladonia vareschii** Ahti, Ann. Bot. Fenn. 23: 218. 1986. Type:
Venezuela, Bolívar, Dist. Piar, Auyántepui, El Peñon, 1600 m alt.,
Vareschi 8704 (holotype VEN, isotype H).   – Fig. 55

Primary thallus not seen. Podetia (3-)5-8(-15) cm tall, 1.0-1.6(-2.2) mm
thick (branches thinner), of indeterminate growth, pale yellow, more or
less variegated with brown over most of its length, browned at the tips,
necrotic parts darkening but not really melanotic, moderately branched,
but main stems usually distinct within 1-2 mm from the tips and branchlets
short (0.2-1 cm); branching type anisotomous dichotomy and trichotomy,
more rarely tetrachotomy; axils usually perforated; tips acute, straight,
divergent; squamules absent. Podetial surface matt to somewhat shiny,

Fig. 55. *Cladonia vareschii* (Sipman 40337 (B)). Bar = 2 cm.

near tips continuously corticate, lower down becoming shallowly areolate, revealing the translucent or browned stereome, so that the general habit is often clearly variegate; usually esquamose, rarely sparsely squamulose; esorediate. Podetial wall 200-320 μm thick; cortex 12-15 μm; medulla 100-140 μm; stereome 100-180 μm, sharply delimited, horny; surface of central canal glossy, somewhat striate. Conidiomata terminal, 200-280 × 120 μm, ovoid, containing purple slime. Hymenial discs up to 2 mm wide, dark brown. Chemistry: usnic and thamnolic acid as major substances (TLC of 16 specimens). Colour reactions: P+ yellow-orange or P-, K+ yellow or K-, KC+ yellow.

D i s t r i b u t i o n   a n d   e c o l o g y : Endemic to the Guiana Shield in Colombia, Venezuela and Guyana, occurring on the lowland white sands and on rock outcrops and in acidic bogs from 85 to 2700 m alt. It seems to be common on the flat tops of many tepuis in Venezuelan Guayana. In the Guianas limited to the sandstone tableland on the Pakaraima Mts. of Guyana. Here it grows in scrub savanna on open places, from 400 to 1200 m alt. Over 50 collections studied (GU: 27).

S e l e c t e d   s p e c i m e n s : Exsicc.: Lichenotheca Latinoamericana 118. Guyana: Upper Mazaruni Distr., 8 km N of Kamarang, Mt. Latipu, Sipman & Aptroot 19148 (B); Karowtipu Mt., Boom & Gopaul 7639 (BSG, NY); Region Potaro-Siparuni, Kaieteur Falls National Park, DePriest 9274 (H); Region 7 (Upper Mazaruni Distr.), N of Paruima Mission, Aymatoi savanna, Sipman 39923 (B, BRG).

N o t e s : *Cladonia vareschii* is a very distinctive species, characterized by greenish-yellow colour (usnic acid) and robust podetia with a strong, horny stereome, short branchlets, open axils and variegate basal parts. By its strongly developed main stems it resembles most closely *C. sufflata* and *C. steyermarkii*. The first frequently grows together with it, but can be distinguished by its soft stereome and the central canal showing a pruinose surface. The second has a thick cortex and a pruinose central canal surface and lacks a variegated colour pattern on the outside. Outside of the *Unciales* group *C. flavocrispata* can be very similar, and is best distinguished by the presence of (sometimes scarce) squamules.
The closest relative may be *Cladonia spinea*, which differs mainly in one conspicuous feature: predominant isotomous branching instead of anisotomous branching. This difference holds the suggestion that *C. vareschii* might be an aged stage of *C. spinea*. In the field both appeared to be easily distinguishable, however.

45. Cladonia variegata Ahti, Lichenologist 22: 266. 1990. Type: Venezuela, Bolívar, Dist. Piar, Macizo del Chimantá, Churí-tepui, 2250 m alt., Ahti *et al.* 44912 (holotype VEN, isotypes B, BM, COL, DUKE, H, MYF, NY, US). – Fig. 56

Primary thallus unknown. Podetia 3-9 cm tall, of indeterminate growth, ash-grey to (often) dark brown, often forming distinct, 2-4 cm wide, semiglobose heads, or flat-topped cushions up to 20 cm in diam.; branching type very dense, irregular, isotomous dichotomy (rarely trichotomy or tetrachotomy); no main stems differentiated; internodes short, usually 1-2 mm; axils closed or with small perforations, rarely widely open; branches 0.3-0.5(-0.8) mm wide, broadly divergent, extreme tips acute, straight (giving a spiny appearance to the lichen). Podetial surface sometimes uniformly pale grey, but usually appearing variegate, with 0.1-0.2 mm wide, white verruculae, containing algal glomerules, scattered in brown to white interspace, which is thinly arachnoid on very young branchlets but with naked stereome through most of the podetia; squamules very rare (only in DePriest 9271), minute, rounded or slightly crenulate, c. 0.2-0.3 mm long and wide. Podetial wall 80-140 μm thick; cortex absent (or with very thin corticoid layer on algal glomerules); medulla (including the algal layer) 60-80 μm or absent; stereome 80-120 μm, horny, fragile; surface of central canal glossy. Conidiomata terminal, 200 × 100 μm, containing red slime. Hymenial discs not seen. Chemistry: thamnolic acid (TLC of 6 specimens). Colour reactions: P+ yellow, K+ yellow, KC-.

Fig. 56. *Cladonia variegata* (Sipman 40341 (B)). Bar = 2 cm

Distribution and ecology: Known mainly from the Guayana Highlands in Venezuela, where it is locally common in paramoid vegetation on hummocks of wet, oligotrophic bogs from c. 1000 to 2250 m alt. Probably more widespread in the Guayana Shield, recently found in Colombia. In the Guianas, occasional in the sandstone tableland of the Pakaraima Mts., in scrub savanna on open spots, where it grows terrestrial on rock surfaces which remain long wet by water trickling out of surrounding vegetation patches after rain, from 400 to 1200 m alt. Over 25 collections studied (GU: 14).

Selected specimens: Exsicc.: Lichenotheca Latinoamericana 120. Guyana: Region Potaro-Siparuni, Kaieteur Falls National Park, DePriest 9271 (H); Region 7 (Upper Mazaruni Distr.), N of Paruima Mission, Aymatoi savanna, Sipman 39926 (B, BRG).

Notes: *Cladonia variegata* is often associated with *C. pulviniformis*, a similar, densely branched species, which is usually more grey in colour, forms narrower, rounded heads, and is softer in texture because of the absence of cartilaginous stereome.
The rather dull appearance of this species makes it easily overlooked, or mistaken for dead thalli.
The full chemical spectrum, not yet observed on Guianas material, includes thamnolic and occasionally barbatic acids as major substances, decarboxythamnolic acid, 4-0-demethylbarbatic acid and unknown compounds as minor substances.

EXCLUDED SPECIES

Cladonia corymbites Nyl. was erroneously reported for the Guianas (Sipman 2007) and, therefore, the species should be excluded from our flora area.

120

ACKNOWLEDGEMENTS

A special word of thanks is for the organisers of expeditions to Guyana which kindly allowed us to participate, notably Ben ter Welle (Utrecht, 1985, 1992) and Paula T. DePriest (Washington, 1996, 1997). They enabled us to collect essential material for this study. Grants from the Academy of Finland and other sources in the 1980's and 1990's, as well from the Smithsonian Institution Guyanan Biodiversity Program in 1996 and 1997 enabled us to do herbarium and field work in Guyana and Venezuelan Guayana. Much appreciated is also the manyfold help from our colleagues, such as Soili Stenroos, Otto Huber and S.S. Tillett, and the curators of the numerous herbaria studied (mainly B, BRG, FH, H, NY, TUR, U, US, VEN).

NUMERICAL LIST OF ACCEPTED TAXA

(Unnumbered taxa are reported from the Guayana Highlands, not from
the Guianas)

1. Cladia Nyl.
    1-1. C. aggregata (Sw.) Nyl.
    1-2. C. globosa Ahti

2. Cladonia P. Browne
    2-1.   C. argentea (Ahti) Ahti & DePriest
           C. atrans (Ahti) Ahti & DePriest
    2-2.   C. cartilaginea Müll. Arg.
    2-3.   C. cayennensis Ahti & Sipman
    2-4.   C. ceratophylla (Sw.) Spreng.
           C. chimantae Ahti
    2-5.   C. confusa R. Sant.
    2-6.   C. corallifera (Kunze) Nyl.
           C. crassiuscula Ahti
    2-7.   C. crustacea Ahti
    2-8.   C. dendroides (Abbayes) Ahti
    2-9.   C. densissima (Ahti) Ahti & DePriest
    2-10. C. didyma (Fée) Vain.
           C. flavocrispata Ahti & Sipman
    2-11. C. furfuraceoides Ahti & Sipman
           C. granulosa (Vain.) Ahti
    2-12. C. guianensis S. Stenroos
           C. hians Ahti
           C. huberi Ahti
    2-13. C. hypoxantha Tuck.
    2-14. C. isidiifera Ahti & Sipman
    2-15. C. maasii Ahti & Sipman
    2-16. C. meridionalis Vain.
    2-17. C. miniata G. Mey.
    2-18. C. mollis Ahti & Sipman
    2-19. C. peltastica (Nyl.) Müll. Arg.
    2-20. C. persphacelata Sipman & Ahti
    2-21. C. pityrophylla Nyl.
    2-22. C. polyscypha Ahti & L. Xavier
    2-23. C. polystomata Ahti & Sipman

2-24. C. prancei Ahti

2-25. C. pulviniformis Ahti

C. rangiferina (L.) F.H. Wigg. subsp. abbayesii (Ahti) Ahti & DePriest

C. rappii A. Evans

2-26. C. recta Ahti & Sipman

2-27. C. rotundata Ahti

2-28. C. rugulosa Ahti

2-29. C. rupununii Ahti & Sipman

2-30. C. secundana Nyl.

2-31. C. signata (Eschw.) Vain.

2-32. C. sipmanii Ahti

2-33. C. spinea Ahti

2-34. C. sprucei Ahti

2-35. C. steyermarkii Ahti

2-36. C. subdelicatula Vain. ex Asahina

2-37. C. subradiata (Vain.) Sandst.

2-38. C. subreticulata Ahti

2-39. C. subsphacelata Sipman & Ahti

2-40. C. subsquamosa Kremp.

2-41. C. substellata Vain.

2-42. C. sufflata Ahti

2-43. C. termitarum Ahti

2-44. C. vareschii Ahti

2-45. C. variegata Ahti

COLLECTIONS STUDIED
(Collection numbers in **bold** refer to types)

GUYANA

Abraham, A.A., 123 (2-15)
Ahti, T. *et al.*, 52899 (2-34); 52900 (2-18); 52901 (2-30); 52902 (2-32); 52908 (2-32); 52909 (2-32); **52910** (2-18); 52912 (2-18); 52913 (2-22); 52914 (2-6); 52915 (2-22); 52917 (2-37); 52918 (2-11); 52919 (2-37); 52921 (2-30); 52922 (2-30); 52924 (2-32); 52925 (2-32); 52927 (2-32); 52928 (2-8); **52929** (2-32); 52932 (2-33); 52933 (2-32); 52934 (2-8); 52935 (2-22 cf.); 52935 (2-32); 52936 (2-8); 52936 (2-32); 52937 (2-11); 52939 (2-32); 52955 (2-44); 52956 (2-25); 52957 (2-33); 52958 (2-25); 52959 (2-22); 52982 (2-21); 52983 (2-22); 52989 (2-32); 52990 (2-8); 52994 (2-28); 52996 (2-25); 52997 (2-25); 52999 (2-21); 53003 (2-32); 53004 (2-45); 53010 (2-11); 53014 (2-25); 53015 (2-40); 53018 (2-6); 53020 (2-43); 53021 (2-22); **53023** (2-43); 53025 (2-11); 53035 (2-43); 53036 (2-16); 53038 (2-16); 53039 (2-44); 53040 (2-10); 53042 (2-41); 53043 (2-33); 53045 (2-5); 53046 (2-25); 53047 (2-33); 53048 (2-16); 53048 a (2-39); 53052 (2-24); 53054 (2-36); 53060 (2-22); 53098 (2-33); 53102 (2-11); 53105 (2-22); 53161 (2-19); 53162 (2-9); 53163 (2-25); 53164 (2-31); 53172 (2-34); 53173 (2-33); 53178 (2-8); 53186 (2-6); 53187 (2-24); 53192 (2-8); 53333 (2-36); 53352 (2-13); 53354 (2-5); 53357a (2-24); 53358a (2-37)

Appun, C.F., 406 (2-34)
Bartlett, A.W., 8041 (2-19)
Bleij, E. & J.C. Biesmeijer, s.n. (2-22)
Boom, B. *et al.*, 7135 (2-6); 7244 (2-24); 7248 (2-11); 7250 (2-10); 7543 (2-31); 7598 (2-2); 7613 (2-38); 7623 (2-23); 7639 (2-44); 7640 (2-4); 7663 (2-6); 7665 (2-19); 7673 (1-1)
Cooper, A., 53 (2-33)
Cornelissen, J.H.C. & H. ter Steege, C-4 (2-32); C-81 (2-10); C-88 (2-22); C-114 (2-22); C-115 (2-22); C-126 (2-22); C-176 (2-22); C-180 (2-10); C-181 (2-22); C-181 a (2-6); C-184 (2-22); C-214 b (2-24); C-214 (2-22); C-216 a (2-22); C-401 (2-22); C-402 (2-22)
Cruz, J.S. de la, 1834 (2-31); 1835 (2-32); 1836 pp (2-19); 1836 pp (2-32); 2056 (2-31); 2057 (2-32); 3514 (2-37); 3415 (2-19); 3415 pp (2-31); 3527 (2-10); 3547 (2-32); s.n. (2-22)
DePriest, P.T. *et al.*, 9013 (2-37); 9014 (2-37); 9015 (2-37); 9016 (2-37); 9017 (2-37); 9018 (2-37); 9019 (2-37); 9051 (2-37); 9096 (2-22); 9097 (2-22); 9150 (2-37); 9154 (2-37); 9179 (2-18); 9190 (2-22); 9198 (2-37); 9199 (2-40); 9200 (2-37); 9201 (2-37); 9202 (2-37); 9203 (2-10); 9204 (2-37); 9205 (2-37); 9216 (2-31); 9217 (2-32); 9218 (2-32); 9219 (2-32); 9220 (2-32); 9222 (2-32); 9223 (2-19); 9224 (2-34); 9225 (2-34); 9226 (2-32); 9227 pp (2-32); 9227 pp

(2-33); 9227 pp (2-41); 9228 (2-22); 9229 (2-11); 9230 (2-6); 9231 (2-22); 9233 (2-22); 9238 (2-30); 9241 (2-22); 9250 (2-25); 9251 (2-25); 9252 (2-25); 9254 (2-33); 9255 (2-5); 9256 (2-31); 9257 (2-31); 9258 (2-31); 9259 (2-31); 9260 (2-32); 9261 (2-1); 9262 (2-34); 9263 (2-34); 9264 (2-34); 9265 (2-34); 9266 (2-25); 9267 (2-25); 9270 (2-19); 9271 (2-45); 9272 (2-25); 9273 (1-1); 9274 (2-44); 9275 (2-44); 9276 (2-33); 9277 (2-44); 9278 (2-33); 9279 (2-33); 9280 (2-33); 9281 (2-33); 9282 (2-44); 9283 (2-19); 9284 (2-19); 9285 (2-19); 9286 (2-19); 9287 (2-19); 9290 (1-1); 9291 (2-28); 9292 (2-28); 9293 (2-28); 9294 (2-22); 9295 (2-22); 9296 (2-22); 9297 (2-22); 9298 (2-22); 9304 (2-22); 9305 (2-22); 9306 (2-22); 9311 (2-22); 9312 (2-30); 9316 (2-18); 9317 (2-16); 9329 (2-22); 9330 (2-19); 9332 (2-18); 9361 (2-18); 9373 (2-31); 9374 (2-30); 9375 (2-39); 9376 (2-22); 9378 (2-22); 9380 (2-22); 9387 (2-28); 9388 (1-1); 9391 (2-18); 9392 (2-44); 9393 (2-33); 9394 (2-28); 9400 (2-28); 9406 (2-44); 9407 (2-22); 9408 (2-10); 9422 (2-32); 9423 (2-19); 9425 (2-31); 9430 (2-37); 9433 (2-11); 10001 (2-18); 10004 (2-41); 10005 (2-30); 10007 (2-21); 10008 (2-21); 10011 (2-33); 10012 (2-31); 10013 (2-31); 10014 (2-25); 10015 (2-45); 10016 (2-33); 10017 (2-25); 10018 (2-34); 10019 (2-44); 10020 (2-9); 10021 (2-19); 10022 (2-9); 10023 (2-25); 10024 (2-45); 10025 (2-11); 10026 (2-10); 10028 (2-19); 10030 (2-25); 10031 (2-11); 10032 (2-6); 10034 (2-28); 10036 (2-34); 10037 (2-19); 10038 (2-31); 10039 (2-5); 10040 (2-34); 10041 (2-25); 10042 (2-45); 10043 (2-19); 10044 (2-34); 10045 (2-25); 10046 (2-25); 10047 (2-31); 10048 (2-8); 10049 (2-33); 10051 (2-33); 10052 (2-6); 10053 (2-22); 10054 (2-37); 10056 (2-19); 10057 (2-1); 10058 (2-10); 10059 (2-8); 10061 (2-44); 10062 (2-22); 10063 (2-16); 10064 (2-19); 10068 (2-11); 10069 (2-19); 10070 (2-25); 10071 (2-25); 10072 (2-45); 10418 (2-36); 10431 (2-11); 10461 (2-27); 10462 (2-25); 10463 (2-31); 10465 (2-34); 10466 (2-34); 10467 (1-2); 10468 (2-31); 10469 (2-9); 10470 (2-30); 10471 a (2-6); 10471 b (2-33); 10472 b (2-25); 10473 (2-45); 10474 (2-9); 10475 (2-33); 10476 (2-38); 10477 (2-1); 10478 (2-28); 10480 (2-35); 10481 (2-42); 10482 (1-1); 10483 (2-34); 10484 (2-1); 10485 (2-8); 10486 (2-34); 10487 (2-42)

Görts-van Rijn, A.R.A. *et al.*, 476 (2-34); 476 a (2-19)

Hahn, W.J. *et al.*, 3851 pp (2-33); 4134 (2-30); 4570 (2-33); 4682 (2-30)

Hellum, A.K., s.n. (2-33)

Henkel, T.W. *et al.*, 6 (2-28); 8 (2-34); 9 (2-38); 11 (2-22); 3052 (2-38)

Hoffman, B. *et al.*, 1586 (2-11); 1696 (2-38); 3347 (2-32); 3347 A (2-33); 3419 a (2-22)

Jansen-Jacobs, M.J. *et al.*, 838 (2-30); 879 pp (2-30); 879 pp (2-34); 893 (2-4); 3266 (2-6); 3444 (2-4); 4871 (2-22); 5420 (2-19)

(1-1); 18671 (2-14); 18792 (2-19); 18825 (1-1); 18872 (1-1); 18884 (1-1); 18897 (2-6 cf.); 18901 (2-10); 18976 (2-10); 18989 (2-37); 18994 (2-21); 19061 (2-17); 19129 (2-34); 19130 (2-28); 19131 (2-6); 19132 (1-1); 19133 (2-21); 19134 (1-1); 19135 (2-30); 19136 (2-21); 19137 (2-23); 19138 (2-23); 19139 (2-12) ; 19140 (2-19); 19141 (2-11); 19141 a (2-22); 19142 (2-30); 19143 (2-32); 19144 (2-28); 19145 (2-4); 19146 (2-6); 19147 (2-11); 19148 (2-32); **19149** (2-20); 19149 a (2-19); 19150 (2-19); 19151 (2-25); 19152 (2-32); 19153 (1-2); 19154 (2-31 thick); 19155 (2-38); 19156 (2-27); 19157 (2-5); 19158 (2-9); 19159 (2-8 cf.); 19160 (2-41); 19161 (2-42); 19162 (2-25); 19163 (2-34); 19164 (2-15 cf.); 19177 (2-19); 19178 (2-6); 19179 (2-22); 19242 (2-19); 19243 (2-5); 19243 b (2-32); 19244 (2-34); 19244 b (2-22); 19248 (2-6); 19251 (2-22); 19251 b (2-11); 19251 c (2-23); 19258 (2-31); 19263 (2-32); 19264 (2-30); 19321 (2-19); 19322 (2-20); 19323 (2-19); 19324 (2-19); 19325 (2-30); 19327 (2-31); 19328 (2-31); 19329 (2-25); 19330 (2-27); 19330 a (2-5); 19330 b (2-19); 19331 (2-31); 19332 (2-19); 19333 (2-19); 19334 (2-9); 19335 (B, H) (2-28); 19336 (2-19); 19338 (2-21); 19338 b (2-11); 19338 c (2-21); 19340 (2-6); 19370 (2-21); 19372 (2-10); 19379 (2-22); 19387 (2-34); 19398 (2-40); 19467 (2-16); 19494 (2-20); 19496 (2-28); 19497 (2-10); 19497 a (2-22); 19498 pp (2-19); 19498 pp (2-33); 19499 (2-11); 19499 a (2-37); 19499 b (2-40); 19500 (2-

6); 19501 (2-32); 19502 (2-11); 19577 (2-22); 39774 (2-30); 39815 (2-28); 39838 (2-40); 39839 (2-30); 39860 (2-20); 39861 (2-19); 39862 (2-6); 39865 (2-28); 39867 (2-11); 39868 (2-37); 39870 (2-35); 39871 (2-9); 39872 (2-38); 39873 (2-25); 39874 (2-1); 39875 (2-38); 39876 (2-34); 39877 (2-32); 39878 (2-27); 39878 A (2-9); 39879 (2-44); 39880 (2-33); 39881 (2-33); 39882 (2-25); 39883 (2-9); 39885 (1-1); 39886 (2-9); 39887 (2-38); 39888 a (2-31); 39888 (2-9); 39889 (2-9); 39889 A (2-9); 39890 (2-5); 39891 (2-34); 39892 (2-31); 39893 (2-38); 39894 (2-34); 39895 (2-33); 39896 (2-41); 39897 (2-28); 39898 (1-2); 39901 (2-38); 39902 (2-25); 39903 (2-25); 39904 (1-2); 39905 (2-35); 39906 (2-35); 39907 (2-28); 39908 (2-44); 39910 (2-19); 39914 (2-25); 39915 (2-45); 39916 (2-25); 39917 (2-34); 39918 (2-34); 39919 (2-21); 39920 (2-45); 39921 (2-33); 39922 (2-42); 39923 (2-44); 39924 (2-9); 39925 (2-41); 39926 (2-45); 39927 (2-33); 39928 (2-33); 39929 (2-1); 39930 (2-44); 39931 (2-42); 39932 (2-44); 39933 (2-41); 39934 (2-44); 39935 (2-34); 39936 (2-41); 39937 (1-2); 39938 (2-35); 39939 (2-38); 39947 (2-4); 39969 (2-32); 40286 (1-1); 40287 (1-1); 40288 (2-1); 40289 (2-9); 40290 (2-34); 40291 (2-34); 40292 (2-34); 40293 (2-6); 40294 (2-41); 40295 (2-32); 40296 (2-32); 40297 (2-32); 40298 (2-32); 40299 (2-33 cf.); 40300 (2-33 cf.); 40301 (2-11); 40302 (2-16); 40303 (2-18); 40304 (2-19); 40305 (2-19); 40306 (2-19); 40307 (2-21);

40308 (2-22); 40309 (2-22); 40310 (2-25); 40311 (2-25); 40312 (2-19); 40313 (2-25); 40314 (2-25); 40315 (2-25); 40316 (2-30); 40317 (2-28); 40318 (2-28); 40319 (2-30); 40320 (2-30); 40321 (2-31); 40322 (2-45); 40323 (2-31); 40324 (2-31); 40325 (2-32); 40326 (2-39); 40327 (2-33); 40328 (2-33); 40329 (2-33); 40330 (2-33); 40331 (2-33); 40332 (2-33); 40333 (2-33); 40334 (2-41); 40335 (2-19); 40336 (2-33 cf.); 40337 (2-44); 40338 (2-44); 40339 (2-44); 40340 (2-44); 40341 (2-45); 40347 (2-33); 40348 (2-33); 40349 (2-8); 40350 (2-44); 40351 (2-33); 40352 (2-33); 40353 (2-25); 40354 (2-33); 40355 (2-33); 40356 (2-45); 40357 (2-25); 40358 (2-34); 40359 (2-28); 40360; (2-44); 40361 (2-1); 40362 (2-31); 40363 (2-33); 40364 (2-5); 40365 (2-8); 40366 (2-41); 40367 (2-19); 40368 (2-45); 40369 (2-34); 40447 (2-20); 41284 (2-0); 41288 (2-0); 41497 (2-37); 56967 (2-22); 57028 (2-22); 57126 (2-19); 57127 (2-26); 57128 (2-19); 57129 (2-8); 57130 (2-31); 57131 (2-24); 57132 (2-26); 57133 (2-6); 57134 (2-26); 57135 (2-22); 57136 (2-32); 57153 (2-24); 57154 (2-40); 57155 (2-32); 57156 (2-33); 57157 (2-11); 57158 (2-22); 57749 (2-29)

Smith, A.C., 2177 pp (2-9); 2177 pp (2-19); 2177 pp (2-30); 2177 pp (2-33); 3200 (2-4); 3631 (2-4); 3654 (2-11); 3654 pp (2-37)

Stenroos, S., 4789 (2-22); 4794 (2-11); 4799 (2-23); 4802 (2-40); 4812 b (2-30); 4831 (2-28); 4854 (2-27); 4876 (2-31); 4881 (2-44); 4897 (2-8); 4899 (2-9); 4916 (2-35); 4918 (2-1); 4922 (2-44); 5876 (2-6)

Stoffers, A.L. et al., 2 (2-30); 3 (2-6); 5 (2-32); 5 a (2-32)

U.G. Bio, 6 (2-33); 90 (2-33); 106 (2-34)

SURINAME

Allen, B., 19307 (2-22)

Aptroot, A., 14823 (2-37); 14829 (2-3); 14854 (2-37); 14961 (2-22); 14962 (2-25); 14963 (2-6); 14964 (2-6); 14965 (2-6); 14966 (2-37); 14967 (2-19)

Arnoldo, M., 3485 (2-22)

Benjamins, H.D., s.n. (2-10); s.n. (2-19); s.n. (2-37)

Donselaar, J. van, 2768 (2-31)

Florschütz, P.A. et al., s.n. (2-4); 176 (B 60 0163714, H, L) (2-2); 656 (2-22); 759 pp (2-25); 759 pp (2-27); 764 (2-10); 765 pp (2-25); 765 pp (2-27); 766 (2-19); 772 (2-31); 839 (2-6); 1285 (2-37); 2030 (2-37); 2963 (2-4); 2969 (1-1); 4571 (2-6); 4576 (2-10); 4740 (2-22); 4791 (2-37); 4791 A (2-10); 4792 (2-6); 4797 E (2-37).

Focke, H.C., 850 (2-37)

Geijskes, D.C., s.n. (2-37)

Heinsdijk, D., 56 (2-19); 73 (2-19)

Herrnhut, s.n. (2-6)

Irwin, H.S. et al., 55523 (2-6)

Jansen-Jacobs, M.J. et al., 6172 (2-37); 6178 (2-6); 6180 (2-6); 6727 (2-37); 6784 b (2-31)

Kegel, H.A.H., 166 (2-6); 996 (2-25); 997 (2-6); 1416 (2-6); 1448 (2-24)

Kramer, K.U. & W.H.A. Hekking, 2655 (2-6); s.n. (2-6)

INDEX TO SYNONYMS, NAMES IN NOTES AND UNCERTAIN RECORDS

Cenomyce
corallifera Kunze = 2-6
sphaerulifera Taylor = 2-10
Cladina
argentea Ahti = 2-1
atrans Ahti = Cladonia atrans
confusa (R. Sant.) Follmann & Ahti = 2-5
dendroides (Abbayes) Ahti = 2-8
densissima Ahti = 2-9
densissima Ahti fo. *decolorans* Ahti, see 2-9, note
peltastica Nyl. = 2-19
rangiferina subsp. *abbayesii* (Ahti) W.L. Culb. = rangiferina subsp.
    abbayesii; see 2-34, note
rotundata (Ahti) Ahti = 2-27
sandstedei (Abbayes) Ahti, see 2-8, note
sprucei (Ahti) Ahti = 2-34
Cladonia
aggregata (Sw.) Spreng. = 1-1
ahtii S. Stenroos, see 2-3, note
bacillaris (Ach.) Genth, see 2-10, note
brasiliensis (Nyl.) Vain., see 2-6, note
capitellata (Hook. f. & Taylor) C. Bab. fo. *interhiascens* (Nyl.) Vain.,
    see 2-19, note
carassensis Vain. (Sipman & Aptroot 1992: 92), see 2-32, note
caribaea S. Stenroos, see 2-14, note
carnea Hampe (Schomburgk 1849: 1041) = 2-6
carneobadia Hampe (Schomburgk 1849: 862, 1041): unknown status
ciliata Stirt., see 2-9, note
coccifera (L.) Willd., see 2-6, note
coccifera (L.) Willd. subsp. *hypoxantha* (Tuck) Vain = 2-13
coccifera (L.) Willd. var. *stemmatina* Ach., see 2-6
coccinea Hampe (Schomburgk 1848: 1041): nom. nud., probably for
    2-6
cocomia Hampe (Schomburgk 1848: 862, 1041): misspelling, see *C.*
    coccinea
confusa fo. *bicolor* (Müll. Arg.) Ahti, see 2-5, note
coniocraea (Flörke) Spreng., see 2-37, note
connexa Vain. (Sandstede 1939: 101, map 79): very probable
    misidentification
corallifera var. *kunzeana* Vain. = 2-6
corallifera fo. *kunzeana* Vain. = 2-6

corymbites Nyl., excluded from the area, see 2-2, note

crispatula (Nyl.) Ahti, see 2-32, note

cubana (Vain.) A. Evans (Evans 1947: 39 and 1955: 100): very probable misidentification

dactylota Tuck.: see Cladoniaceae, notes (6)

didyma var. *vulcanica* (Zoll. & Moritzi) Vain. = 2-10

digitata (L.) Schaer. (Vainio 1887: 123; Sandstede 1932: 70): very probable misidentification

dimorphoclada Robbins, see 2-41, note

ecmocyna G. Mey., see 2-19, note

"ecmozyma Hampe", see 2-19, note (probable misspelling)

erythromelana Müll. Arg. = 2-30

fallax Abbayes, see 2-5, note

fimbriata (L.) Fr. var. *apolepta* "(Ach.) Vain." (Hue 1898: 276): perhaps refers to 2-22

fimbriata [var.] *chondroidea* [subvar.] *subradiata* Vain. = 2-37

furfuracea Vain., see 2-11, note

labradorica Ahti & Brodo, see 2-41, note

macilenta Hoffm., see 2-10, note

medusina (Bory) Nyl. var. *submedusina* (Müll. Arg.) Vain. ex Zahlbr. = 2-19

miniata G. Mey. var. *anaemica* (Nyl.) Vain., see 2-30, note

miniata var. *erythromelaena* (Müll. Arg.) Zahlbr. = 2-30

miniata [fo.] *secundana* (Nyl.) Vain. = 2-30

miniata var. *secundana* (Nyl.) Zahlbr. = 2-30

muscigena Eschw. = 2-10

"muscigera Eschw. var. *pulchella* Tuck." = 2-10

mutabilis Vain., see 2-28, note

pityrea (Flörke) Fr. var. *ramosa* Flot. ex Hampe, see 2-22, note

pityrea fo. *cladomorpha* Flörke, see 2-22, note

pityrophylla [var.] α. *spruceana* Vain. = 2-21

pityrophylla fo. *spruceana* Vain. = 2-21

polytypa Vain., see 2-28, note

pyxidata (L.) Hoffm., see Cladoniaceae, notes (6)

pyxidata (L.) Fr. var. *chlorophaea* (Flörke ex Sommerf.) Flörke (Vainio 1894: 232), see 2-30, note

ramulosa (With.) J.R. Laundon, see 2-22, note

rangiferina var. *abbayesii* Ahti = *Cladonia rangiferina* subsp. *abbayesii* (expected subspecies)

rangiferina [var.] *signata* Eschw. = 2-31

rangiferina (L.) F.H. Wigg. subsp. *abbayesii* (Ahti) Ahti & DePriest (expected subspecies)

rangiformis Hoffm. (Meyer 1818: 297): very probable misidentification

reticulata (J.L. Russell) Vain., see 2-38, note

reticulata fo. *lacunosa* (Tuck.) Vain., see 2-38
salzmannii Nyl. fo. *ascypha* Abbayes = 2-32
sandstedei Abbayes, see 2-8, note
sandstedei fo. *dendroides* Abbayes = 2-8
squamosa Hoffm., see 2-40, note
stellaris (Opiz) Pouzar & Vĕzda, see 2-9, note
stenroosiae Ahti, see 2-41, note
strepsilis (Ach.) Grognot (des Abbayes 1956: 265): in need of
    confirmation
submedusina Müll. Arg. = 2-19
subsquamosa (Nyl. ex Leighton) Crombie, see 2-40, note
subsquamosa [var.] ß. *granulosa* Vain. = *Cladonia granulosa* (expected
    species)
subtenuis (Abbayes) A. Evans fo. *subtenuis* (Ahti 1961: 70), refers to
    2-31
subulata (L.) F.H. Wigg., see 2, type
uncialis (L.) F.H. Wigg., see 2-35, note
verticillata Hoffm. var. *cervicornis* (Ach.) Flörke (Vainio 1894: 191),
    unknown status
vicaria R. Sant. (des Abbayes 1961: 117): very probable
    misidentification
vulcanica Zoll. & Moritzi = 2-10
vulcanica fo. *isidioclada* (Mont. & Bosch) Abbayes = 2-10
Clathrina
    aggregata (Sw.) Müll. Arg. = 1-1
Lichen
    aggregatus Sw. = 1-1
    ceratophyllus Sw. = 2-4
    subulatus L., see 2, type
Scyphophorus
    didymus Fée = 2-10

Alphabetic list of families of series E – Lichens
occurring in the Guianas

Family delimitations according to the status of the year 2011. Those published are followed by the family name valid at the time of publication, chronological fascicle number and year.

Agyriaceae
Arthoniaceae
Arthopyreniaceae
Asterothyriaceae
Biatorellaceae
Brigantiaeaceae
Caliciaceae
Candelariaceae
Catillariaceae
Celotheliaceae
Chrysothricaceae
Cladoniaceae
Clavariaceae
Coccocarpiaceae
Coenogoniaceae
Collemataceae
Crocyniaceae
Fuscideaceae
Gomphillaceae
Graphidaceae
Gyalectaceae
Haematommataceae
Hygrophoraceae
Icmadophilaceae
Lecanoraceae
Lichinaceae
Lobariaceae
Lyrommataceae
Massalongiaceae
Megalariaceae
Megalosporaceae
Megasporaceae
Melaspileaceae

Microtheliopsidaceae
Monoblastiaceae
Mycocaliciaceae
Mycoporaceae
Naetrocymbaceae
Ochrolechiaceae
Pannariaceae
Parmeliaceae
Peltulaceae
Pertusariaceae
Phlyctidaceae
Physciaceae (Pyxinaceae)
1. 1987
Pilocarpaceae
Porinaceae (Trichotheliaceae)
2. 1993
Psoraceae
Pyrenothricaceae
Pyrenulaceae
Ramalinaceae
Roccellaceae
Sphinctrinaceae
Stereocaulaceae
Strigulaceae
Teloschistaceae
Tephromelataceae
Thelenellaceae
Trypetheliaceae
Verrucariaceae

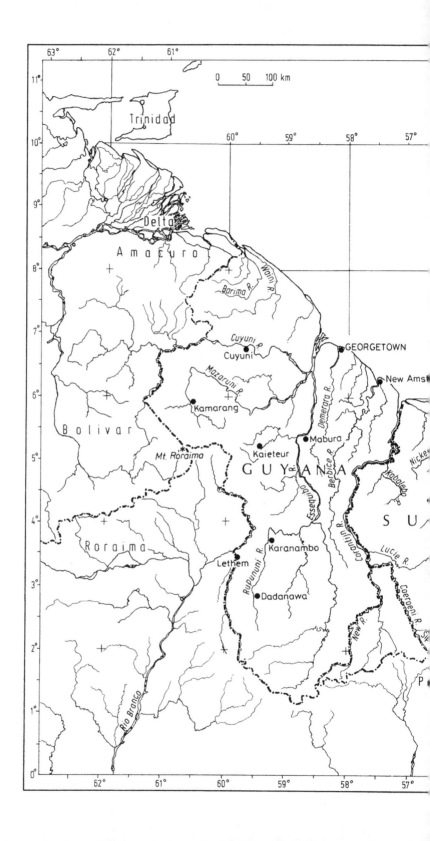

THE GUIANAS
uyana, Suriname, French Guiana

55° 54° 53°
10°
9°
8°

7°
6°
5°
4°
3°
2°
1°
0°

MARIBO
Commewijne R.
Zanderij
kopondo
Brownsberg
Suriname R.
Yca R.
St. Laurent
Kourou
CAYENNE
St. Elie
Marowijne R.
Mana R.
Sinnamary R.
Approuague R.
NAME
F R E N C H
G U I A N A
St. Georges
Tapanahony
Inini R.
Saül
Oelemari R.
Litani R.
Marouini R.
Tampoc R.
Camopi R.
Oyapock R.
A m a p á

55° 54° 53° 52° 51° 50°